ENVIRONMENT AND MAN
VOLUME SEVEN

Measuring and Monitoring the Environment

ENVIRONMENT AND MAN: VOLUME SEVEN

Titles in this Series

Volume 1 *Energy Resources and the Environment*

Volume 2 *Food, Agriculture and the Environment*

Volume 3 *Health and the Environment*

Volume 4 *Reclamation*

Volume 5 *The Marine Environment*

Volume 6 *The Chemical Environment*

Volume 7 *Measuring and Monitoring the Environment*

Volume 8 *The Built Environment*

ENVIRONMENT AND MAN
VOLUME SEVEN

Measuring and Monitoring the Environment

General Editors

John Lenihan
O.B.E., M.Sc., Ph.D., C.Eng., F.I.E.E., F.Inst.P., F.R.S.E.

Director of the Department of Clinical Physics and Bio-Engineering, West of Scotland Health Boards, Professor of Clinical Physics, University of Glasgow.

and

William W Fletcher
B.Sc., Ph.D., F.L.S., F.I.Biol., F.R.S.E.

Professor of Biology and Past Dean of the School of Biological Sciences, University of Strathclyde; Chairman of the Scottish Branch of the Institute of Biology; President of the Botanical Society of Edinburgh.

1978

ACADEMIC PRESS·NEW YORK & SAN FRANCISCO
A Subsidiary of Harcourt Brace Jovanovich, Publishers

Blackie & Son Limited
Bishopbriggs
Glasgow G64 2NZ

© 1978 Blackie & Son Ltd.
First published 1978

All rights reserved.
No part of this publication may be reproduced,
stored in a retrieval system, or transmitted,
in any form or by any means,
electronic, mechanical, recording or otherwise,
without prior permission of the Publishers

International Standard Book Number

0-12-443507-6

Library of Congress Catalog Card Number

75-37435

Printed in Great Britain by
Thomson Litho Ltd., East Kilbride, Scotland

Background to Authors

Environment and Man: Volume Seven

KENNETH MELLANBY, C.B.E., Sc.D. was the first Director of Monks Wood Experimental Station, and is Honorary Professorial Fellow at the University College of South Wales and Visiting Professor of Biology at the University of Leicester. He is the author of many scientific papers and books concerned with the environment, and is Editor of the International Journal *Environmental Pollution*. He is a regular contributor to *Nature*.

MICHAEL J. SAXBY, B.Sc., Ph.D., F.R.I.C., is a Research Supervisor at the Leatherhead Food Research Association. He has held a number of posts in Britain and abroad in the food and tobacco industries.

D. B. JAMES, C.Chem., Dip. Tech., D.L.C., M.R.I.C., F.I.W.E.S., is Chief Chemist and Bacteriologist at the Newcastle and Gateshead Water Company. In 1975 he was awarded the Gans Medal of the Society for Water Treatment and Examination for research into bacterial pollution of water by gulls.

JOHN LENIHAN, O.B.E., M.Sc., Ph.D., C.Eng., F.I.E.E., F.Inst.P., F.R.S.E. has been involved since 1953 in the application of novel analytical techniques to the study of clinical, environmental, historical and forensic problems. He is Editor of *Advances in Activation Analysis* and a member of the Editorial Board of the *Journal of Radioanalytical Chemistry*.

JOHN P. GRIFFIN, B.Sc., Ph.D., M.B., B.S., L.R.C.P., M.R.C.S., is Senior Principal Medical Officer in the Medicines Division of the Department of Health and Social Security, and Honorary Consultant Physician at the Lister Hospital, Stevenage. From 1961 to 1971 he was Head of Clinical Research, Riker Laboratories.

Series Foreword

MAN IS A DISCOVERING ANIMAL—SCIENCE IN THE SEVENTEENTH CENTURY, scenery in the nineteenth and now the environment. In the heyday of Victorian technology—indeed until quite recently—the environment was seen as a boundless cornucopia, to be enjoyed, plundered and re-arranged for profit.

Today many thoughtful people see the environment as a limited resource, with conservation as the influence restraining consumption. Some go further, foretelling large-scale starvation and pollution unless we turn back the clock and adopt a simpler way of life.

Extreme views—whether exuberant or gloomy—are more easily propagated, but the middle way, based on reason rather than emotion, is a better guide for future action. This series of books presents an authoritative explanation and discussion of a wide range of problems related to the environment, at a level suitable for practitioners and students in science, engineering, medicine, administration and planning. For the increasing numbers of teachers and students involved in degree and diploma courses in environmental science the series should be particularly useful, and for members of the general public willing to make a modest intellectual effort, it will be found to present a thoroughly readable account of the problems underlying the interactions between man and his environment.

Preface

"SCIENCE IS MEASUREMENT," SAID HELMHOLTZ. SOME MAY FEEL THAT this is an over-simplication, but it must be constantly borne in mind by those working in the environmental field. We need less emotionalism and more measurement; it is only by measurement that we can gauge when things are going wrong and take appropriate steps to put them right. This volume is a valuable one in that it indicates some of the ways in which some of the variables in an environment are monitored.

Professor Kenneth Mellanby writes about *biological indicators*—plants and animals whose presence or absence, health or disease, reveal much of the nature of surrounding conditions. In general, biological indicators may not indicate the precise nature of the polluting substances, but many are quite remarkably sensitive to pollutants and are ideal early-warning systems.

Some of the by-products of our very successful technology can produce chemicals with particularly nasty side-effects, and it is important that we should have the means of detecting them if they find their way into our foodstuffs. Dr. Michael J. Saxby deals with some of the sophisticated techniques that have been developed to ensure that even minute quantities of contaminants such as pesticides, polychlorinated biphenyls, dioxins and vinyl chloride can be detected. It is important to realize that some dangerous chemicals can arise through natural processes. Dr. Saxby does well to remind us that some of them, e.g. aflatoxins (which are produced by the attack of certain fungi on certain plants) are among the most toxic substances known to man. Their presence must be carefully monitored too.

Our single most important naturally-occurring commodity is water, but the twin developments of technology and population growth have meant that many water supplies in the world have become heavily polluted by effluents. Now, according to Mr. D. B. James, standards are being laid down to control water quality at various stages in the water cycle in order to reverse a trend which, in developed countries, had almost reached disaster level. This chapter deals with many of the methods used in these control systems and with the significance of the findings.

Human hair is of great interest to students of the environment. Professor John Lenihan points out that it is very durable; is easily stored and transported; its proteins have an affinity for many heavy metals; only small samples are required for analysis; it records contamination from inside the body and from outside; and it acts as an integrating dosemeter over its period of growth. Techniques have been developed which can detect trace elements below the level of 1 ppm. Arsenic, mercury, lead, are among the metals discussed in this chapter.

It is not only Man's external environment that must be carefully monitored. His internal environment is at least as important. Recent years have seen a considerable expansion in the use of drugs, and while most of the activity has been beneficial in that the death rate from infections has been cut dramatically, pain has been lessened, and many psychological illnesses have been alleviated or controlled, we must be constantly on the look-out for side effects. In his chapter on "Drug Toxicity" Dr. John P. Griffin deals with adverse reactions to drugs that are in common pharmaceutical use.

W. W. F.

Contents

CHAPTER ONE— BIOLOGICAL METHODS OF
ENVIRONMENTAL MONITORING   1
by K. Mellanby

> Introduction. Advantages and limitations of biological indicators. Lichens. Mosses. Fungi. Vascular plants. Ecological investigations of pollution. Conclusion. Further reading.

CHAPTER TWO— ANALYSIS OF FOOD   14
by M. J. Saxby

> Introduction. Analysis of foodstuffs for compounds resulting from environmental pollution. Pesticides. Insecticides. Organochlorine insecticide residues. Organophosphorus insecticide residues. Herbicides. Fumigants. Fungicides. Polychlorinated biphenyls. Polychlorodibenzo-p-dioxins. Heavy metals. Vinyl chloride. **Analysis of naturally formed compounds in food.** Nitrosamines. Mycotoxins (i) aflatoxins, (ii) other mycotoxins. Polycyclic hydrocarbons. **Food additives.** Food dyes and antioxidants. **Conclusions.** Further reading.

CHAPTER THREE— SOME ASPECTS OF WATER QUALITY
CONTROL   by D. B. James   37

> Introduction. The Water Cycle. Contamination. Pollution. **Measurements necessary for control.** Chemical Analyses. Titrimetric or volumetric analysis. Colorimetry. Emission Spectroscopy. Gravimetric analysis. Specific electrodes. Odour and taste. Radioactivity. Microbiological examination of water. The coliform organism. Multiple-tube test. Membrane filtration. Bacteria growing in agar. Viruses in water. Biological examination. The Utermöhl method. Membrane filter method. **Significance of water analysis.** Organoleptic factors. Physico-chemical factors. Undesirable or toxic factors. Biological factors. Microbiological factors. **Scope of water analysis.** Public water supplies. Recreational use of water. Aquatic life. Industrial use of water. **Monitoring of water quality. Control of water quality.** Rain water. Lakes. Rivers. Underground water. The provision of a safe water supply. Further reading.

CHAPTER FOUR— HAIR AS A MIRROR OF THE
ENVIRONMENT   by J. M. A. Lenihan   66

Growth and structure of hair. Metals in hair. Analysis of hair. Activation analysis. Atomic absorption. Arsenic and smoking. Arsenic in detergents. Occupational poisoning. Attempted suicide. Mercury hazards in dental practice. Environmental lead. Environment and disease. The historical environment. Further reading.

CHAPTER FIVE— DRUG TOXICITY   87
by J. P. Griffin

The magnitude of the problem. The gathering of data. Some of the drugs involved. The nature of the problem. **Inherent drug toxicity.** The thalidomide problem. The continual appearance of new hazards and new dimensions of adverse reactions—diethylstilboestrol and adenocarcinoma of the vagina. Phenacetin renal damage—a problem of toxicity and drug abuse. **Toxicity due to drug interaction.** Drug absorption interference. Plasma protein binding. Drug metabolism. Renal excretion. Receptor sites. **The patient factor in drug toxicity.** Genetic factors. Fast and slow acetylation of drugs. Familial abnormal response to neuromuscular blocking drugs. Porphyria. The cutaneous porphyrias. Acute porphyrias. Mixed porphyrias. Drug-induced haemolytic anaemia in patients with inherited erythrocyte abnormalities. African type. Mediterranean type. Other common types of severe G6PD deficiency. Participation of genetic factors in thromboembolism in women on oral contraceptive therapy. Regional factors in adverse reactions to drugs. Age factors. Disease states. Renal failure. Liver disease. Myasthenia gravis. Gout and diabetes. Abnormal eyes (shallow anterior chamber). Hypersensitive and allergic reactions to drugs. Drug-induced allergic skin reactions. Allergic contact dermatitis. Skin rashes from systematically administered drugs. Drug-induced allergic asthma. Allergic drug-induced haemolytic anaemia. Drug regulatory authorities. **Application for clinical trial certificate, Application for marketing or product licence. Conclusion.** Further reading.

INDEX   129

CHAPTER ONE

BIOLOGICAL METHODS OF ENVIRONMENTAL MONITORING

KENNETH MELLANBY

Introduction

Living animals and plants are affected by the chemical and physical conditions of their environment, and their reactions may be used to measure some of these conditions. In this way living organisms may act as "biological indicators". The term may be used in a number of different ways. To the ecologist, a *biological indicator* may be defined as *a species of plant or animal characteristic of climatic, soil and other conditions in a particular region or habitat*. Thus an indicator species may be one which occurs naturally under restricted conditions, and its presence may be diagnostic of such factors as soil acidity, the absence of frost, or the presence of ample supplies of water. In such cases the indicator is denoting conditions favourable to its establishment and growth. However, it should be noted that in many cases plants thought to be indicators are not actually favoured by local conditions, but are simply more resistant to some factor or factors which prevent the establishment of other species. Thus roses are often thought to "prefer" a clay soil; in fact they grow rather better on other formations. Other garden flowers dislike clay and so we get the impression that the tolerant roses have a preference for these conditions. Here the missing species are the real biological indicators.

Even the most inexperienced amateur naturalist recognizes differences in the vegetation and the animals present in different localities, and that these differences are related to temperature, moisture and various other conditions. If he sees a field containing rushes and willow trees, he knows that the ground will be moist. Even an inexperienced gardener knows that rhododendrons and heather indicate that the soil in which they are flourishing is acid. By looking at a photograph of an area of natural vegetation, a botanist can usually infer a good deal about the soil, the climate, and even the latitude and longitude of the area where the picture was taken. We all rely on biological information when we try to describe and quantify our environment.

Advantages and limitations of biological indicators

• The term *biological indicator* may be used in a more precise way to describe those living organisms which are sensitive to specific substances, or to specific environmental conditions, and which react in an easily recognizable way. In particular, the use of these organisms in the study of pollution will be discussed.

The environment is considered to be polluted when some sort of damage can be recognized. Material objects, stone or paint, can be affected by chemicals, and these effects may be measured and quantified. However, harmful effects on living organisms may occur when much lower concentrations of some pollutants are present, and more accurate estimates of the quantitative effects of the different pollutants may be possible. To the biologist, the use of biological indicators is attractive. He is concerned with the way in which pollution may harm man and other animals and, if a plant or animal is involved in measuring levels of pollution, he can relate this to the way in which less-sensitive organisms are affected. If a sensitive indicator is unharmed, conditions are likely to be "safe" for others.

The most familiar biological indicator is the canary, used to detect carbon monoxide in coal mines. As long as the canary appears to be healthy, the level of this gas must be well below that harmful to the miners at work alongside; for, with its small body size and high metabolic rate, the canary is rendered unconscious very quickly. When this happens, the miners have plenty of time to escape the danger of asphyxiation.

Biological indicators have both advantages and disadvantages over chemical methods of measuring pollutants. As already mentioned, effects on them can be related to effects on other living organisms, and they clearly indicate something in the environment which can be thought of as being "harmful". However, they cannot always indicate the precise nature of the polluting substance, and results are affected by other environmental factors. In many coal mines, the commonest risk is from carbon monoxide and, when a canary falls over unconscious, it is most likely that carbon monoxide is to blame. But the simple reaction of the bird could have been caused by all manner of other poisons and conditions, by ill health or metabolic upsets, or even, in the immortal words of W. S. Gilbert, by

a rather tough worm in its little inside.

In fact biological indicators are really only of much use if we already know what the most important pollutant is likely to be.

The situation is complicated even further by the way that environmental conditions affect the reactions of the indicators. Thus a well-nourished plant may sometimes be more, and sometimes less, susceptible to certain forms of damage than is one grown in an impoverished soil. The atmospheric humidity can be important; some toxic gases do little damage unless the air is moist. Temperature is obviously important. Cold-blooded animals, and many plants, may have a very low metabolic rate on a cool day, and pollution may then do little damage. On the other hand, in warm weather a fast-growing plant may be less susceptible than is the same species when the growth rate is slowed by the cold.

Biological indicators are generally organisms which are more than usually sensitive to specific substances, but none is wholly specific. Many react in a similar manner to different chemicals, and there may also be synergistic effects caused by the inter-reaction of two substances (either or both of which may, alone, be less toxic). As already mentioned, their great advantage is that they tend to indicate that the environment is, in some way, unhealthy for them, and therefore that conditions may be approaching those where other plants, animals, or even man may be in danger. Only occasionally will indicator species supply data easily transformable into accurate numerical levels of polluting substances. Biological indicators are therefore best thought of as useful for investigating the environment, but only as one method which usually requires to be supplemented by chemical and physical analyses of particular substances and conditions. They may often, like the canary, be used as an early-warning system to show that there is a risk of something "going wrong".

So far in this chapter only those plants and animals which are adversely affected, usually by low concentrations of toxic substances, have been considered as biological indicators; but other processes in plants and animals may also be used to monitor pollutants. One of the most important of these is the ability of some organisms to concentrate substances within their tissues, thus making it possible for us to detect these substances which may otherwise be present at levels where accurate analysis is difficult. The process of "concentration along food chains" is well known and, though it has sometimes been misunderstood and exaggerated, it is true that some predatory birds and mammals at the top of the food pyramid do selectivly retain substances, e.g. fat-soluble organochlorine insecticides, in some of their tissues. This concentration may give rise to pathological symptoms or even death in the predators but, more often, apparently harmless but easily analysed levels allow these creatures to be used to estimate environmental levels of certain chemicals.

Fish and some other aquatic animals selectivly extract and concentrate, sometimes by factors of tens of thousands, some of the chemicals in the water in which they live. A terrestrial animal which breathes air will pick up only small amounts. A water-breathing animal comes into intimate contact with large volumes of water in its efforts to obtain enough oxygen; and the more polluted the water, the lower the level of oxygen one normally finds, and so the larger the volume which has to be "breathed". (Where these chemicals do no obvious damage, they may again be estimated by analysing suitable tissues.)

Dead plant materials used to monitor pollution levels could be considered to come within the scope of this chapter. Moss bags—nylon bags containing sphagnum moss—are widely used to collect lead and other heavy metals which are taken up from the air by the moss and can then be analysed. It is also possible to use synthetic plastics such as polythene for this purpose—clearly not a "biological" process. Also one finds that some indicator plants which are easily damaged continue, after death, to absorb differentially the substances which killed them. Can the post-mortem findings still be considered "biological"?

A great many different plants and animals have been used as biological indicators. It is not intended to give a comprehensive account of all such, but rather to describe the main mechanisms which have proved useful to man.

Lichens

Lichens have been used as biological indicators of air pollution to a greater extent than any other group of plants. These organisms consist of a fungus and an alga (or, occasionally, more than one species of alga) living in an intimate and symbiotic relationship. They obtain their nutrients either from dry deposition on their surfaces, from rain and, of course, by the photosynthetic activities of the algal component. Some lichens are exceedingly susceptible to damage by air pollutants, particularly sulphur dioxide and fluorine. This has been known for more than a hundred years, and large areas of urban and industrial development are lichen-free.

Many different pollutants are discharged into the atmosphere by man, from his dwellings, his factories and his machines. In Britain, soot and smoke, once the most serious discharges, have been greatly reduced in quantity, largely because little raw coal is now burned by industry or in houses in urban areas. This improvement in air quality has been brought about mainly by economic forces (oil, with its low labour costs replacing coal as an industrial fuel) plus the Clean Air Act of 1964. The largest amount of any pollution now is that arising from sulphur dioxide, of which

BIOLOGICAL METHODS OF ENVIRONMENTAL MONITORING

some six million tonnes are discharged annually in Britain, with similar or higher levels in other industrialized countries. Although levels of sulphur dioxide are highest nearest to their sources, the gas is also widely distributed in differing concentrations over the world, some being known to pass over oceans to give rise to international pollution problems.

Fluorine, discharged into the atmosphere by industry, particularly from brickworks and aluminium smelters, is a serious but usually local pollutant, dangerous levels seldom occurring more than a few kilometres from the source.

The distribution of various different species of lichens has been used to map sulphur dioxide levels over wide areas, and fluorine near to sources of emission. Although some work has been done on the physiological effects of the pollutants, the presence or absence of selected species with different susceptibility has been the main subject of study. Investigations of the rate of growth (or of regression) have also been made around new sources of pollution such as recently erected aluminium smelters.

Studies have been made at various levels of sophistication. Dr D. L. Hawksworth of the Commonwealth Mycological Institute in London and his co-workers, and Dr O. L. Gilbert of Sheffield University, England, have published information showing how a wide range of species of lichen responds to different mean levels of sulphur dioxide. They have shown that these plants are, species by species, remarkably consistent in the levels of sulphur which they are able to tolerate. (Some 85 species have been studied in depth.) Very dry conditions may prevent some lichens from developing, and wet areas in polluted zones sometimes contain species for which the air might be expected to be slightly too sulphurous. There is no evidence that short periods of high sulphur levels are seriously damaging to lichens; in this they appear to differ from some vascular plants.

There is no doubt that the expert lichenologist, able to identify accurately the species he finds, is able to use these plants to the greatest advantage in studying air pollution. Detailed investigations of small areas with considerable gradients of sulphur or (more likely) fluoride levels really require such expertise. However, useful studies can be made at a much lower level.

In 1972 the Advisory Centre for Education, Cambridge, England, and the London newspaper the *Sunday Times* collaborated to organize a survey which they hoped would be performed, voluntarily and during the school summer holidays, by British children. The children were invited to write and buy a "Clean Air Research Pack" which consisted essentially of material to enable the children to map different sorts (not, in essence, species) of lichen with different susceptibility to atmospheric pollution. Many thousands of children responded, and a substantial number sent in surprisingly well executed maps dealing with their local conditions.

Although information was given which would have allowed a number of species to be identified, zones were more simply characterized. Thus zone 6 ("really clean air") was where shrubby lichens were likely to be prominent. The simple point made was that "the more it (the lichen) sticks out, the cleaner the air". The most-polluted areas (zone 0) were expected to be free of lichens, the next (zone 1) had only the very-resistant and easily-identified *Lecanora conizaeoides*. In zone 2 leafy lichens began to appear, and increased and became more vigorous in zones 3 and 4.

The results were centrally analysed and collated. They enabled a map to be drawn for the whole of Britain. It was interesting to note that it agreed very closely with a map for sulphur dioxide levels based on the analytical survey carried out by the Warren Spring Laboratory, and with the results obtained using accurate lichen species determinations by Dr D. L. Hawksworth and Dr F. Rose. A number of similar simple surveys done by local authorities and other groups have been equally rewarding.

It must be stressed that the fact that lichen surveys can be used in this way does not in any way detract from the more detailed work done by the expert lichenologists. Lichens can serve as indicators in several different ways, and at various levels of sophistication.

Incidentally, it is interesting to note how some susceptible lichens are damaged by very low levels of sulphur dioxide in the air. The most affected species are eliminated by only $30\,\mu g/m^3$, and many leafy species cannot exist where SO_2 levels exceed $50\mu g/m^3$. Such levels do not return sufficient sulphur to the soil (if it is sulphur-deficient) to encourage the healthy growth of many grasses and cereal crops, and levels perhaps a hundred-fold higher are needed to cause discomfort to human beings.

Recently aluminium smelters have been erected in areas, for instance in Anglesey in Wales, where the air was previously very pure, and where shrubby and leafy lichens abounded. The smelters have, unfortunately, discharged a great deal of fluorine into the surrounding countryside. The lichens which might have been expected to be susceptible have in fact regressed and died. They have also been shown, by analysis, to absorb a great deal of fluorine into their bodies. It is not always easy to tell exactly when a lichen dies, but it is clear that the susceptible plants are killed by their exposure. However, they can still be used to monitor the pollution, for the dead thalli continue to absorb fluorine, and levels reach several thousand parts per million.

Susceptible lichens are severely affected by sulphur dioxide and hydrogen fluoride. They can also play some part in monitoring other pollutants. They absorb, possibly passively, considerable amounts of heavy metals, e.g. lead from the exhaust gases of automobiles. Little work has been done on damage to plants from this source; few leafy or foliose lichens are generally

BIOLOGICAL METHODS OF ENVIRONMENTAL MONITORING

present where traffic is dense, probably because this occurs in urban areas where sulphur dioxide is also abundant.

Lichens have also been used in surveys of long-life nuclides (e.g. caesium-137 and strontium-90) arising from nuclear explosions. Unexpectedly high levels have been found in the lichens in the arctic tundra; this has demonstrated the aerial distribution pattern of the nuclides in the northern hemisphere. There is no evidence to suggest that the lichens have been adversely affected by the levels of radioactive materials they have taken up; they have simply acted as biological concentrators of the nuclides.

Mosses

Mosses are said to be about as sensitive to air pollution as are lichens, and the centres of most cities are free of mosses. A number of research workers have found that mosses are affected by many pollutants, including sulphur, fluorine, ozone and heavy metals, and some pollution maps have been drawn with the aid of these plants. However, for some reason they have been used less than lichens, and little information about the exact relationship to different pollution levels of different species of moss exists. No doubt their value as biological indicators will be exploited more in the future, if they are found to show advantages for this purpose.

Living mosses accumulate heavy metals. This process may, up to certain limits, seem harmless to the plants, but some mosses are damaged, for instance by low levels of cadmium and zinc. This damage seems to have been little exploited as a measure of pollution levels; it is not very easy to assess visually the degree of damage, and to give it a numerical value.

The ability to accumulate heavy metals has been used, and mosses exposed to different pollution levels show, on analyses, repeatable differences in their metal content. However, as was stated earlier, mosses have been mostly used in so-called "moss bags" where dead dried moss (usually sphagnum) is exposed and then analysed. The moss is clearly a good substance for this purpose, but it is not a biological process, nor can the dead moss be considered to be a biological indicator. Other pollutants, e.g. sulphur and various particulates, can be collected on plastic netting in the same way.

Fungi

When air pollution, including ground-level concentration of sulphur dioxide, was worse in British cities than it is today, urban gardeners had one advantage over their rural counterparts. City roses were free from the disease known as *black spot*, caused by the fungus *Diplocarpus rosae*. Today the levels of sulphur dioxide have fallen, and the fungus has returned to city parks and gardens. It has thus shown itself to be an efficient biological indicator, rampant only when the level of sulphur dioxide falls below $100\,\mu g/m^3$.

As a rule fungi are more easily damaged by a number of sulphur compounds (some of which may be used as fungicides) than are vascular plants. The city roses free from black spot were unaffected by the sulphur dioxide which cured the disease. Air pollution can cause economic damage to mushroom crops at levels where other plants are unharmed. It might therefore be expected that fungi could be widely used to measure some forms of air pollution.

In fact the higher fungi, i.e. those which produce mushrooms and toadstools, have been little used as indicators. This is perhaps because these fruiting bodies tend to appear somewhat erratically, so that systematic studies are difficult. No doubt pollution affects the subterranean mycelium, but this is even more difficult to study. It seems unlikely therefore that these fungi will be widely exploited as indicators of pollution in the near future.

Fungi causing disease in higher plants seem easier to use. The case of black spot has already been mentioned, but an even more promising species is *Rhytisma acerinum*, the cause of tar spot disease in the sycamore. The symptoms are large black blotches on the leaves of this tree. They are absent when the average concentration of sulphur dioxide is above $90\,\mu g/m^3$. However, this fungus may be used to indicate lower levels of pollution. Sycamore leaves may be seriously or only slightly affected, and a "tar spot index", based on the number of blotches, can be determined and associated with a particular level of pollution. Maps, similar to those produced for lichens, can then be drawn with the aid of this disease. It has one advantage over lichens in that, as the fungus develops on the leaf of a deciduous tree during the spring and summer, the measurements can be related to atmospheric conditions during a shorter period than may be necessary to affect lichen distribution.

Systematic studies have not been made with other species of disease-causing fungi, but there are numerous observations suggesting their absence from polluted areas, and the possibility of using them as indicators, particularly of sulphur dioxide levels. Some species also appear to be very sensitive to ozone. It seems likely, therefore, that these organisms will be more widely studied in the future.

Vascular plants

Many of the higher plants are susceptible to damage by a variety of air pollutants. This property has often been first noticed because of damage to

BIOLOGICAL METHODS OF ENVIRONMENTAL MONITORING

a crop of economic importance. Sometimes it is found that one cultivar is much more easily damaged than another; the sensitive cultivar is obviously a candidate for use as a biological indicator.

Some whole groups of plants are known to be harmed by comparatively low levels, particularly of sulphur and fluorine. Thus it has long been known that coniferous trees do not flourish in cities, while some deciduous species, for instance the London plane, are much less easily affected. These are thus rather crude indicators of the environmental conditions.

In recent years, however, sophisticated work has been done in this field. Much of this is summarized in the *Pictorial Atlas*, published by the Air Pollution Control Association of Pittsburgh, Pennsylvania, USA. This contains a great number of coloured plates showing typical damage to a host of susceptible plants caused by the more important forms of air pollution. The volume contains, for instance, a list of 20 crop plants, and of 20 trees and shrubs, which are relatively sensitive to ozone. The form that damage takes is then illustrated by coloured photographs. Sulphur dioxide, fluoride, nitrogen oxides, "PAN" (peroxyacyl nitrates), and ethylene are treated similarly.

At first sight it may appear that there is a host of plants which can easily be used to monitor air pollution. Unfortunately, the situation is more complicated. Experienced research workers can indeed use many plants in this way, particularly in areas where they know what sort of pollution to expect. Others may find themselves in difficulties. They may be unable to distinguish between plants damaged by disease and those affected by pollution. Although certain lesions may be characteristic of damage caused by, for instance, sulphur dioxide, very similar effects may be produced by other substances.

It is also not unusual to find different plants of the same species growing quite near to one another, with some showing damage and others appearing unharmed. Both groups may be subjected to the same levels of pollution; the difference may be the result of differences in soil, in fertilizer application, in water. To serve their purpose, indicator plants generally need very careful and uniform cultivation.

Although pollution damage to higher plants can, with experience, be recognized, it may not be easy to relate the degree of damage to exact levels or durations of pollution. Plants may react quite quickly. Thus tobacco or beans may show obvious damage (within 48 hours) from exposure for a few hours to levels of below ten parts of ozone per hundred million (of air). It is not difficult to recognize conditions where the ozone levels are too high to allow the normal growth of susceptible plants. But similar damage can be caused by quite a short exposure, sometimes of only a few minutes, to a high level of ozone, and by a much longer exposure to a lower level of contamination by the same gas.

Very few vascular plants can be easily used to give accurate estimates of the way in which the pollution operated. Nevertheless there are many ways in which biological indicators from the higher plants may be used to monitor the air quality. There are, as has already been mentioned, particular cultivars which are much more easily damaged than other plants of the same variety. This can be well illustrated using tobacco *Nicotiana tabacum.* One well-known variety called Bel W 3 is particularly easily damaged by exposure to air containing raised levels of ozone. If plants of this variety show none of the lesions typical of ozone damage in some particular locality, ozone levels (which have often been found to cause serious damage to commercial tobacco crops in many parts of the United States of America) in those regions will be unlikely to harm other varieties of tobacco or, for that matter, of any other crop. Unless the Bel W 3 plants are very seriously damaged, others will probably be safe. (There is some margin of safety.)

Thus many vascular plants can be used as general indicators of "environmental quality". If a series of plots of different plants, with different susceptibilities to several different pollutants, are planted in a number of areas, locations with "high-quality air" (where no damage can be observed) will be identified and distinguished from more-polluted places. The exact nature of the pollution, the concentration of the gases responsible, and the duration of pollution incidents may not be determined, but real potentialities for damage will be identified. This is the great virtue of the biological indicator.

As mentioned more than once above, only the expert can extract the maximum results from the use of indicator plants, but they can also be used as educative tools in schools and colleges. Professor W. A. Feder and his colleagues at the University of Massachusetts have produced some useful pamphlets, issued by the University's Extension Service. These show how tobacco, beans, gladioli, and other plants can be used to demonstrate the presence of air pollution, particularly that caused by ozone and sulphur dioxide.

Vascular plants, like mosses and lichens, can be used to collect and concentrate some pollutants, when analysis of the tissues may make it possible to evaluate the conditions where the plants grow. Plants already damaged by pollution may be used, but many results have been obtained from apparently unharmed plants which have picked up quite large amounts of substances such as lead emitted in automobile exhausts. Often the plants are merely acting as physical surfaces on which the pollutants are deposited, and so have no specific biological function; but there may also be active absorption by the plant tissues, and effects on the cells and on enzyme processes, in which case exact ambient levels may, with experience, be determined.

There is no doubt that much more research on the effects of pollutants on higher plants remains to be done. These organisms are likely to be increasingly used as improved and more accurate techniques are evolved relating their responses more closely to levels and intensities of atmospheric pollution.

Ecological investigations of pollution

The ecologist can determine a great deal about the physical and chemical conditions of any area by observing the types of plants and animals which are present. He can study pollution in the same way. In a river, for instance, the flora and fauna are profoundly affected by pollution.

So far we have considered only plants as biological indicators. They are used, and have been studied, more than have animals, for two main reasons. First, plants are not mobile. Once a plant is growing in one spot, it remains there even if the conditions deteriorate. If it is susceptible, it may react in a characteristic way to a newly introduced pollutant, in the ways already described. If an animal receives the same treatment, it will probably simply move away in search of more favourable conditions. Secondly, experimenters have been somewhat reluctant to expose living animals to toxic substances; the exposure of higher animals ("vivisection") is, of course, strictly controlled in most countries.

Nevertheless some animals may be useful as biological indicators. Trout and salmon are only found in clean rivers. Many so-called "coarse" fish co-exist with the salmon and trout but, as they are more tolerant of pollution, they are also found in dirtier waters which cannot support the less-tolerant trout and salmon. The most-polluted rivers have no fish at all. Twenty years ago the Thames, flowing through London, was fishless. Today it supports substantial numbers of the more-tolerant species; they "indicate" how conditions have improved. If conditions improve further, this may be demonstrated by the return of trout (from upstream where the water is less polluted) and salmon (from the sea).

Many invertebrate animals can also be used to evaluate the condition of fresh water. As in other cases, such studies are most effective when carried out by experts, but useful results can even be obtained by children. A survey, similar to that described on page 5 for lichen distribution, was made of British rivers and streams. A kit was prepared by the Advisory Centre for Education, and distributed through the *Sunday Times* newspaper. A series of invertebrate animals was described, with photographs and line drawings; these were chosen as being common in very clean water (e.g. stonefly nymphs), tolerant of slight pollution (e.g. caddis fly larva), able to live in highly polluted conditions (e.g. sludge worm). A series including these and other easily identified species enabled children to divide streams into five categories of increasing pollution. Many thousands of children took part in the study, and produced local maps which were combined into a report covering the whole of Britain. The results agreed closely with those produced by experts employed by the water authorities, but added useful local details, e.g. stretches of pollution in otherwise clean rivers, caused by small factories or inefficient sewage plants. One value of this survey was that local involvement often gave rise to improvements in the effluents from sources such as those just mentioned.

Investigations of this kind are probably easier to make in fresh water, which covers a restricted area and from which the animals cannot easily escape, than in terrestrial habitats. Nevertheless there are many animals which can be used in non-aquatic environments. Even man himself, by his reactions, can play some part as a biological indicator.

Conclusion

It is clear that biological indicators can measure many different environmental factors, and that they can perform in many different ways. Their measurements may also be related to very different periods of time. Thus if we are concerned with only one environmental pollutant, sulphur dioxide, we can see that different organisms give different types of information. Lichens, by the presence or absence of different species, give a picture which integrates the effects of long-term exposure, perhaps over many years. Fungi, particularly those like the tar spot fungus, which cause plant diseases, give results relating to periods of weeks or, at most, months. Many higher plants react to short episodes where there are high levels of sulphur dioxide in the air, and show their characteristic symptoms in at most a few days. In all these cases we can recognize when harmful levels of the pollutant have been present, and we can make some estimate of the level which has been reached—but more exact estimates are usually possible.

It is important that we should realize the limitations in the use of any particular biological indicator, and that we should choose the correct organism for any particular purpose. It cannot be too often repeated that the most valuable results are usually obtained only when we know what pollutant is most likely to be present. It is also clear that, for a proper understanding of our environment, a combination of all methods—chemical, physical and biological—must be intelligently combined.

FURTHER READING

Bevan, R. J. and Greenhalgh, G. N. (1976), "*Rhytisma acerinum* as a Biological Indicator of Pollution," *Environmental Pollution* **10**, p. 271.
Craker, L. E. and Feder, W. A. (1972), *Measuring Air Pollution with Plants*, Cooperative Extension Service, University of Massachusetts.
Ferry, B. W., Baddeley, M. S. and Hawksworth, D. L. (1973), *Air Pollution and Lichens*, London, The Athlone Press.
Hawksworth, D. L. and Rose, F. (1970), "Qualitative Scale for Estimating Sulphur Dioxide Pollution in England and Wales using Epiphytic Lichens," *Nature*, London **277**, p. 145.
Jacobson, J. S. and Hill, A. C. (1970), *Recognition of Air Pollution Injury to Vegetation: A Pictorial Atlas*, Air Pollution Control Association, Pittsburgh.
Mellanby, K. (1972), *The Biology of Pollution*, London, Edward Arnold.
Mellanby, K. and Gilbert, O. L. (1974), "Pollution Surveys by British Children," *Environmental Pollution* **6**, p. 159.
Webster, C. C. (1967), *The Effects of Air Pollution on Plants and Soil*, Agricultural Research Council, London.

CHAPTER TWO
ANALYSIS OF FOOD

M. J. Saxby

Introduction

It is difficult to balance the benefits of a rising standard of living against its cost in terms of the deterioration of the environment and the quality of life. Advances in technology have brought valuable benefits, but there has been an accompanying increase in pollution.

It must not be thought that pollution caused by modern technology is the only source of undesirable compounds in foods; many natural components of foods have toxic properties. To the analyst, it is immaterial whether an undesirable component is of natural origin, or whether it is present as a result of human intervention. His job is to devise a method by which the component may be detected and its concentration determined. The analyst is therefore presented with two objectives:

1. To identify or confirm the structure of components in the mixture represented by a foodstuff.
2. To determine the concentration of the components in the mixture.

The former task may involve the isolation of a compound from a complex mixture, followed by the application of a range of chemical degradative experiments and spectrometric tests to the component. If an interesting compound is present only in trace quantities, most of the evidence for a proposed structure is likely to come from a mass spectrometer coupled directly to a gas chromatograph, since this is one of the few instruments capable of providing structural information on sub-microgram amounts of organic material. In many instances the analyst is dealing with a compound of known structure, which can be readily synthesized, and it is only

ANALYSIS OF FOOD

necessary for him to compare the properties of the compound in the mixture with an authentic sample for the necessary identification to be made.

A more common task that the analyst is called upon to undertake is the quantitative determination of specified compounds. The following techniques are commonly used:

1. Paper and thin-layer chromatography.
2. Visible and ultraviolet spectrophotometry.
3. Infrared spectrophotometry.
4. Polarography.
5. Gas chromatography, employing detection by flame ionization, thermal conductivity, electron capture and possible element-specific methods.
6. Atomic absorption.

High-pressure liquid chromatography, spectrofluorometry and mass spectrometry may also be available.

When designing an analytical procedure, due consideration must be given to the likely concentration of the compound, and to the properties of the matrix in which it occurs. In past decades the analyst was rarely confronted with the examination of compounds present below a fraction of one percent. Nowadays it is more likely that he will be asked to quantify a compound at the part per million (ppm) level or lower. The concentration of many compounds is often expressed in parts per billion (ppb) and when this is done, it is generally recognized that reference is made to the American billion (10^9) rather than to the British billion (10^{12}).

An analytical procedure may usually be divided into four stages:

1. Extraction of the component from the matrix of the sample.
2. Removal of interfering substances from the extract.
3. Measurement of a convenient property of the compound.
4. Confirmation of the result.

The removal of a substance from a foodstuff will generally depend on volatility or solubility. A volatile component may be removed by:

(i) Steam distillation, either at atmospheric pressure or *in vacuo*.
(ii) Likens-Nickerson extraction, in which steam distillation is accompanied by extraction from the distillate with a volatile immiscible organic solvent.
(iii) Headspace analysis, in which the compound is sufficiently volatile to be present in an equilibrium concentration in the vapour phase.

If the compound under analysis is not particularly volatile, then solvent extraction is the usual procedure. The choice of the particular solvent depends upon the properties of the compound, though hexane, ether or (for non-aqueous mixtures) acetone are commonly-used solvents. Solvent extraction must be followed by selective removal of compounds which may interfere with the quantitative determination at a later stage. Sometimes it is possible to remove unwanted compounds from the mixture by relying on differences of solubilities in mixtures of solvents. Thus, some pesticides may be selectively extracted from fatty food samples by solution in hexane, followed by partition into a more-polar solvent like acetonitrile. Fats largely remain in the hydrocarbon solvent, whereas the more-polar pesticides are partitioned almost quantitatively into the acetonitrile. Sometimes it is possible to purify the extracts by column chromatography and elution with a carefully chosen series of solvents.

Quantitative analysis of the purified extract can be performed by any method which identifies a physical or chemical property of the compound. In its simplest form this may mean measuring the size of a spot on a paper or thin-layer chromatogram. If the compound possesses a strong absorbance in the ultraviolet or visible spectrum, this property may be used—but in environmental analysis neither of these techniques offers great sensitivity or selectivity. Whenever possible the analyst will use gas chromatography, because it is convenient and can be made both selective and sensitive.

The confirmation of the concentration of a compound in a product may be tested by the use of two separate analytical techniques. In gas chromatography it is usual to quote the retention times on several columns of different polarity as confirmatory evidence. In cases where the analyst needs to be absolutely sure that his result is correct, he must resort to mass spectrometry and, although this technique is costly, it is often justified, e.g. when evidence is needed in litigation.

The analyst in a well-equipped laboratory has a large number of analytical techniques available to him, several of which may be capable of being used for a certain determination. The choice will often be influenced by sensitivity, presence of interfering compounds, and availability of equipment, e.g. biphenyl is a relatively non-toxic fungicide which is applied to citrus fruit; small residues are sometimes found on the peel. The analyst looking for biphenyl in the fruit of citrus products can extract it with hydrocarbon solvents, or he can remove it by steam distillation. If the extract is relatively free from interfering compounds, measurement of the optical density of the extract in the ultraviolet region of the spectrum will be sufficient. Alternatively colorimetric methods may be used. Since biphenyl is a simple aromatic hydrocarbon, it can readily be analysed by gas chromatography without fear of decomposition. The compound fluoresces, so that fluorometric methods can also be used. Thin-layer, or even paper, chromatography could be used for analysis and, if none of these, then **polarography** is possible.

Analysis of foodstuffs for compounds resulting from environmental pollution

(a) *Pesticides*

(i) *Insecticides.* There is evidence, both from archaeological sources and from early writings that man has always had to contend with plant diseases and the presence of pests in the cultivation of fruit, vegetables and grain. The Bible has many references to plant pathology, e.g.:

> I have smitten you with blasting and mildew: when your gardens and your vineyards and your fig trees and your olive trees increased, the palmerworm devoured them: yet have ye not returned unto me, saith the Lord. (Amos, 4, 9).

(Blasting means withering; the palmerworm is a caterpillar.)

The Industrial Revolution in Britain brought about a rapid expansion in the production of heavy chemicals, some of which—particularly salts of arsenic, lead and copper—were found to be effective in the control of pests and diseases. During the early twentieth century, organic compounds, particularly those which contain chlorine atoms in their structure, were introduced—a trend which led to the discovery by Müller in 1940 of the insecticidal properties of DDT (dichlorodiphenyltrichloroethane) and then to a whole series of organochlorine compounds with similar properties. Since that date, hundreds of pesticides, many of them designed for specific tasks, have been developed. Owing to the long persistence of organochlorine pesticides in the environment, substances based on organic compounds of phosphorus, sulphur or nitrogen have recently come on to the market.

The analyst has to find methods for the determination of these compounds, and also related methods for the examination of their degradation products. For compounds containing several chlorine atoms, gas chromatography with the electron-capture detector is the method of choice, whilst for those groups containing phosphorus, sulphur or nitrogen atoms, the analyst has element-specific detectors in association with gas chromatography.

Organochlorine insecticide residues. It was noted (p. 15) that a successful method of analysis involves extraction, removal of interfering substances and detection. This sequence is closely followed in the analysis of chlorinated pesticide residues such as lindane, dieldrin and DDT. In the first stage, the food product is extracted into a hydrocarbon solvent such as hexane, the exact details of this step depending on the nature of the foodstuff. If a high percentage of water is present, the addition of acetone aids extraction, whereas fatty products are best dissolved directly into the hexane. In the latter case, the presence of such a large quantity of fat presents problems, since the pesticide is readily soluble in animal and vegetable fats. The presence of chlorine atoms in the molecule of the pesticide, however, gives it some polar properties (i.e. there is an unequal charge distribution in the molecule), which cause it to be soluble in polar solvents. Dimethylformamide and acetonitrile, for example, are almost immiscible with hexane and will extract most of the organochlorine pesticides out of the hexane, whilst leaving the unwanted fat in the hydrocarbon solvent. These solvents are also miscible with water; in the next step, the solvent is diluted with an aqueous solution of sodium sulphate or sodium chloride, whereupon the pesticide residues are forced out of solution. Hexane is almost immiscible with solvents like acetonitrile, but a small quantity is transferred into the acetonitrile which, on dilution with water, comes out of solution bringing the pesticide with it. Thus, the compounds of interest are now to be found in a small volume of hexane, and they are relatively free of fat.

The extract of pesticides, after partitioning against acetonitrile (or other polar solvent), still contains too many alien compounds for satisfactory analysis by gas chromatography. At this stage a further purification can be carried out by chromatography over a column of florisil (an activated magnesium silicate) and by elution with a mixture of solvents (often consisting of petroleum spirit and diethyl ether). Some workers claim to have separated the pesticide residues into groups by careful choice of the eluting solvents, but in practice this is generally unnecessary, since little trouble is experienced in their separation by subsequent gas chromatography.

Gas chromatography provides the analyst with an excellent method for both separating and quantifying compounds which show some degree of volatility. Organochlorine pesticides contain several chlorine atoms per molecule and are very sensitive to the electron-capture detector, so that this is the usual choice for the analyst. Early electron-capture detectors suffered from the disadvantage that they were non-linear in their response, which meant that doubling the amount injected did not yield a two-fold increase in the size of the peak. Fortunately, electron-capture detectors now becoming available exhibit a linear response over a very wide range.

The electron-capture detector is exceptionally sensitive to organochlorine pesticides, being able to detect as little as a few picograms ($1\,\text{pg} = 10^{-12}$ gram) of compounds such as lindane and aldrin; thus it becomes a relatively easy operation to quantify pesticide residue levels in environment samples in the parts per billion range (10^9).

Detectors which are capable of responding to picogram quantities of a compound can be misleading, since only trace amounts of impurities on the surfaces of the glass vessels may be sufficient to give a response which can be mistaken for the compound under analysis. The analyst must be constantly on his guard against reporting the presence of false compounds. Apart from making sure that all apparatus and solvents are rigorously clean and pure, he may wish to confirm his findings; this may be done in a variety of ways. Normally mass spectrometry will detect only nanogram amounts of compounds, so that trace quantities of pesticide residues may be below the threshold concentration to produce a response.

For the confirmation of the presence of organochlorine pesticides, it is usual to degrade them chemically in such a way that they effectively disappear from the mixture, or alternatively they produce a secondary product which, though still giving a response to the gas-chromatographic detector, will appear at a different point on the chromatogram. Progress is also being made towards the development of other gas-chromatographic detectors which respond to the actual presence of chlorine atoms in a molecule, rather than to their electron-capturing properties. Positive identification by a second unrelated detector would, of course, provide good evidence for the presence of a compound.

Organophosphorus insecticide residues. The methods of extraction and purification of these pesticide residues are very similar to those employed for the organochlorine insecticides.

For the actual estimation of the organophosphorus pesticide, gas chromatography is the most widely used method, since it combines a good resolution of multi-component mixtures of residues with a good specificity. The electron-capture detector responds to many organophosphates but, unless the molecule also contains chlorine atoms, the sensitivity is about ten times lower than that exhibited by the chlorinated hydrocarbons.

However, the two detectors most widely used for the examination of organophosphates are the alkali flame ionization and flame photometric detectors. Some constructions of the former detector are sensitive to 10 pg of parathion and will discriminate between phosphates and hydrocarbons by a factor of 10 000.

Not all laboratories are equipped with gas chromatography having specific detectors, so that mention should also be made of alternative methods. Most organophosphates are strong cholinesterase inhibitors; this property may be used to detect and analyse spots on thin-layer chromatograms. Plates are sprayed with a solution of cholinesterase, which is inhibited at zones at which organophosphates are present. The plates are then sprayed with a solution of a substrate for cholinesterase, usually indoxyl acetate. The areas on the plate where there are no phosphates contain indoxyl and indigo, both of which fluoresce under ultraviolet radiation, so that the organophosphate spots appear as dark areas on a bright fluorescent background. Under ideal conditions, which depend upon the nature of the organophosphate, spots of a few nanograms can be detected.

Sometimes it is convenient to analyse a sample for total organophosphates before examining for individual pesticides. If the method can be automated, then samples can be screened to remove those in which the total organophosphate is below the acceptable level for any single pesticide. The method involves extraction of the sample with a solvent, purification by column chromatography, hydrolysis, digestion with mineral acids, and estimation of the liberated phosphate by measurement of the blue colour produced with ammonium molybdate.

(ii) *Herbicides.* The analysis of herbicide residues in foods is less important than the analysis of insecticides. This is because herbicides are designed to be effective in the control of weeds; application tends to be early in the growth of the plant, so that only very small amounts are found in human foodstuffs.

(iii) *Fumigants.* Fumigants may be defined as volatile pesticides which are lethal against bacteria, yeasts, moulds, insects, nematodes and rodents. The compounds may be gaseous at room temperature, or may be derived from volatile liquids or solids. Residues of the fumigants may be physically bound to a component of the food, and they can be released under reduced pressure, elevated temperature or prolonged aeration. However, residues may also be chemically bound, in which case they are not readily released.

Since all fumigants are readily converted to a gaseous state, it is clear that gas chromatography provides one of the most effective methods of analysis. In many cases it is possible to analyse residues of the fumigant by headspace analysis, followed by gas chromatography. In other instances, solvent trapping in pentane, xylene or butanol is preferred. The choice of solvent is based on the relative retention time of the fumigant, so that the solvent is either eluted well before or after the gas being analysed. Thus m-xylene is preferred for fast-eluting gases like methyl bromide, methyl chloride and ethylene oxide, whereas n-butanol is preferred for hydrogen cyanide and phosphine. Pentane has been used for other fumigants like dichlorvos and ethylene dibromide. Sometimes concentration on charcoal followed by solvent extraction may be used, but this method suffers from problems associated with interference from water vapour in the ambient air.

Dichlorvos (Vapona) is an organophosphorus fumigant which is effective against flying insects, and it can be extracted from food products, particularly fatty samples, by steam distillation. A specific phosphorus detector is particularly valuable for its detection by gas chromatography.

Methyl bromide residues are often determined as total inorganic bromide, but this excludes the possibility that methylation may have occurred in the food. Neutron activation analysis, x-ray fluorescence, and ashing procedures yield figures for total inorganic and organic bromine, and make no allowance for natural bromide levels.

Phosphine (hydrogen phosphide) is an effective fumigant and is readily generated by the action of moisture on tablets of aluminium or zinc phosphide. Little or no free phosphine has been found in treated cereals, but only scant information is available on possible reaction products between phosphine and active components of the food.

One of the most important determinations on fumigants in the food industry is concerned with products which have been treated with ethylene oxide. In 1965, the presence of toxic ethylene chlorohydrin was detected in foods which had been fumigated with ethylene oxide. It is formed according to the following scheme by reaction with moist sodium chloride.

$$\underset{O}{CH_2\!-\!CH_2} + NaCl + H_2O \rightarrow \underset{OH\quad Cl}{CH_2\!-\!CH_2} + NaOH$$

Chemical methods for the detection of the chlorohydrin involve steam distillation, hydrolysis of the distillate, and titration of the liberated chloride. Gas chromatography is a more popular method, and the whole procedure is illustrated by reference to the analysis of pepper. Pepper powder mixed with Celite is placed in a chromatographic column, which is eluted with ether. The eluate is gently evaporated and passed down a column of florisil. The second eluate is adjusted to a known volume, and the ethylene chlorohydrin estimated by gas chromatography.

The determination of chlorohydrins in modified starches is also important, where the chlorohydrins themselves are used in the production of hydroxyalkyl starch ethers which are widely used as food thickeners in baby foods and pie fillings. Direct ether extraction of hydroxypropyl starch yields small amounts of the chlorohydrin, but larger amounts are detected after soaking the starch for 24 hours. Since it is known that starch tenaciously retains many compounds, it is likely that chlorohydrins are held tightly by polar matrices. Attempts to remove them by steam distillation are unsuccessful, but they are liberated by alkaline hydrolysis in a pressure bottle. After hydrolysis, the chlorohydrins are readily extracted with ether and analysed by gas chromatography.

(iv) *Fungicides.* Fungicides are compounds used on farm crops as protective agents against the attack of fungi; they include soil fungicides, seed protectants, foliage and fruit protectants and eradicants. Their chemical structures vary considerably, since they include organomercury compounds, dithiocarbamates (products of the reaction between carbon disulphide and amines), halogen compounds, cationic compounds, and a number of others of miscellaneous structure.

Unlike organochlorine and organophosphorus pesticides, no really suitable multi-residue technique has been devised for all fungicides. Some are susceptible to analysis by gas chromatography, but many others are involatile or thermally unstable, and none is conveniently detected by sensitive electron-capture techniques. A method for the separation and identification of nine systemic fungicides by thin-layer chromatography, with visualization under UV radiation or by a colorimetric-spray reagent, has been described, but it cannot be applied to coloured extracts. Since fungicides are so varied in nature, descriptions will be limited to a few major groups and specific compounds.

Organomercury compounds in foods are extracted with a slightly alkaline solution of cysteine in isopropanol, which is then washed with ether. The mercurials are partitioned into a very dilute solution of dithizone in ether, which is dried and analysed by gas or thin-layer chromatography. Mercury compounds, particularly methyl mercury derivatives, represent an exception to the rule and are very sensitive to an electron-capture detector. Since organomercury compounds are rather toxic, this analysis is of particular importance in some types of fish and marine oils; the possible presence of methyl mercury in tuna fish was the subject of a newspaper scare some years ago.

Dithiocarbamates constitute a most important group of fungicides which are used to control plant fungus diseases. Residues of these compounds have traditionally been determined colorimetrically by measuring evolved carbon disulphide.

More recently gas chromatography has been used for this determination, with the use of an electron-capture or flame-photometric detector.

Thiabendazole belongs to a group of compounds known as benzimidazoles and was originally developed as an anthelminthic for livestock; it is now widely used as a post-harvest fungicide, particularly on citrus fruits. Extraction with ethyl acetate or dichloromethane is followed by thin-layer chromatography. Spots, located under UV radiation, are extracted with methanol and determined spectrophotometrically at 303 nm or by spectrofluorimetry. The limits of detection by these procedures vary from 0·1 to 3 ppm according to the product being investigated, and to the analytical method which is chosen. Alternative methods of examination include gas chromatography of the trimethylsilyl ether or methyl derivatives with flame ionization or nitrogen-specific detection.

Fentin is the commercial name given to the compound triphenyl tin as the acetate or hydroxide. At relatively high levels it is possible to digest the organic material by wet ashing and to determine the tin colorimetrically with cupferron. At levels below 1 ppm the quantity of food product necessary to give a sufficient colour becomes excessive, and more-sensitive methods are needed. Fentin couples with certain hydroxyflavones to give an intensely fluorescent product with an emission wavelength which is sufficiently far from the emission of the reacting hydroxyflavone to provide a sensitive method of detection.

Biphenyl is a fungicide much used on citrus fruits, and several procedures have been developed for its analysis. In its simplest form, the peel is steam-distilled, extracted with a hydrocarbon solvent like hexane and, after removal of interfering substances, the intensity of the absorption is measured at 248 nm. Colorimetric methods have been developed which rely on the conversion of biphenyl derivatives, through nitration, to coloured products. Fluorometry has also been used. Gas chromatography with flame ionization has also been used where only high-tolerance limits were required. The strong absorption in the ultraviolet region of the spectrum of the biphenyl molecule lends itself well to analysis by high-pressure liquid chromatography with an ultraviolet detector.

(b) Polychlorinated biphenyls

The chemical properties of polychlorinated biphenyls (known generally as PCBs), which make them important industrial chemicals, include their excellent thermal stability and their resistance to strong mineral acids and alkalis. They are insoluble in water and have a low vapour pressure. The chemical structures of the PCBs consist of two joined benzene rings, each of which may include 1 to 5 chlorine atoms. It will be noticed that their structures bear a resemblance to insecticides like p,p'-DDT, and for this reason they are sometimes mistaken for it during analysis.

typical PCB   p,p'-DDT

The major uses of PCBs have been in capacitors, plasticizers, transformer fluids, hydraulic fluids, heat transfer fluids and lubricants. At one time they were also used in carbonless duplicating papers, but this practice has ceased, because the PCBs were becoming incorporated into cardboard, through the waste-paper chain. The viscosity of the PCBs increases with chlorine content of the molecules, and the lower members of the group can be distilled at atmospheric pressure without appreciable decomposition. Since PCBs are so very stable, it is hardly surprising that they are to be found, albeit at low levels, in the environment. As they are very soluble in lipids like body-fat, PCBs tend to become concentrated along the foodchain. A number of examples of poisoning after exposure to PCBs have been reported; probably the best known occurred in Japan in 1968, when more than 900 people were affected after the consumption of contaminated rice oil.

Since PCBs have a degree of volatility and possess a number of chlorine atoms, gas chromatography with electron-capture detection is the obvious method of analysis.

(c) *Polychlorodibenzo-p-dioxins*

During 1957, millions of chickens in the Midwestern and Eastern United States died of a disease which was characterized by excessive fluid in the pericardial sac. Since the component responsible for the disease was at the time unknown, it became known as the chick edema factor (CEF). Similar outbreaks of the disease occurred in later years, when it was discovered that the toxic factor was to be found in the involatile lipid fraction of the feedstuff. The acute toxicity of CEF led to the isolation of a compound, and later a number of compounds, representing a whole series of chlorinated dibenzo-p-dioxins. The presence of these highly chlorinated compounds in the food chain is of concern since they are lipid-soluble non-biodegradable, and tend to become concentrated in fats; many of them are acutely toxic and even carcinogenic.

The source of these compounds has been found to be contaminants of chlorinated phenols, which are widely used in agriculture and industry as bactericides, fungicides, defoliants, herbicides and wood preservatives.

The detection and quantification of these compounds in fats presents no real problems to the analyst, when their properties are fully understood. The fat sample is saponified in alkali, and the non-saponifiable material is eluted over a column of alumina, with an ether-petroleum mixture. Since the p-dioxins are extremely stable compounds, other organic material which remains in the eluate from the alumina column may be removed by washing with concentrated sulphuric acid. The finally purified extract may then be analysed by gas chromatography and detection is by electron capture.

Since the p-dioxins are a family of compounds, it is likely that their presence will be detected as a series of peaks on the chromatogram, and their quantitation may be achieved by a technique similar to that used for the PCBs, in which the sum of all the products of peak heights and retention times are compared with a series of standards.

(d) *Heavy metals*

The chemical control of food products has become more refined in recent years, and in particular the role of inorganic elements and compounds has been recognized. Contamination of foods with heavy metals may make them toxic, and many countries severely limit the levels of many elements which are allowed in foods.

Most methods for trace-metal analysis employ the technique of atomic absorption, which has been developed extensively since the first studies in 1955. Most atomic absorption analyses are performed in flame atomizers, in which the sample is reduced to a fine spray of droplets and introduced into a flame where high-temperature chemical reactions destroy any organic matter and dissociate inorganic compounds into atoms. Atoms of each element absorb light at characteristic wavelengths; by measuring the degree of absorbance at this wavelength, the concentration of the particular element in the sample may be determined and related to the concentration of the element in the original food. More recently, purely thermal methods of atomization have been developed, in which the sample, placed on a carbon rod, is dissociated by electrical heating. Flameless atomization methods are sensitive in the $1/10^9$ region.

Most food samples require pretreatment before analysis. This may be achieved by chemical solubilization or hydrolysis, extraction of the elements of interest, or by destruction of the sample, leaving the inorganic residue for analysis. Simple food samples may require the initial treatment, but complex mixtures of fats, proteins and carbohydrates require acid hydrolysis for 5-10 minutes with about 8N hydrochloric acid.

This treatment is unsuitable for lead and silver owing to poor recoveries. Elements of interest may sometimes be selectively extracted with the chelating reagent, ethenediamine tetra-acetic acid (EDTA). Thus, boiling the foodstuff with 0.1M EDTA solution will extract copper, nickel, iron and chromium at pH 6. Then wet digestion or dry ashing is probably most widely used. Dry ashing needs no reagents; it is suitable for large samples, but it is slow and may lead to losses due to volatilization or absorption onto the crucible walls. Wet digestion on the other hand is generally faster than dry ashing, and losses are reduced; but it requires the use of very pure reagents, needs more operator attention, and is limited to samples weighing less than 10 grams.

Often elements of interest to the food analyst are present only in trace quantities, and the resulting concentration in the prepared sample may be too low for detection. Under these conditions, solvent extraction methods are widely used. An added advantage is obtained in that the sensitivity of many determinations is often increased dramatically in the presence of organic compounds. For many elements, a 1% solution of ammonium pyrrolidine dithiocarbamate chelates the metal into a form which is readily soluble in methyl isobutyl ketone.

Cadmium is an extremely toxic element which, when absorbed from dust or ingested from food, accumulates in the kidneys. Some workers believe that it is one of the most serious environmental pollutants, even at low concentrations; it accumulates particularly in the germ of wheat and rice, and is also concentrated in seafoods. Wet-digestion methods are suitable for detection, and the limit of 0·7 nanogram per millilitre (ng per ml) is adequate for most samples without concentration. The wavelength of the emission is at 228·8 nm and is relatively free from interference.

Chromium is essential for sugar metabolism in the body; for the food analyst, high levels of chromium in a food may suggest contamination from plated cooking vessels.

Because of the low natural levels of chromium in foods, it may be necessary to ash and extract into methyl isobutyl ketone in order to obtain a sufficiently concentrated sample. Some workers consider that hexavalent chromium is more toxic than the trivalent form, since it is more readily absorbed in the stomach but, unless a chemical separation is initially undertaken, analysis by atomic absorption will give only the total chromium content.

Copper is essential to certain enzyme functions, but high doses are toxic. In food analysis, the primary interest is in dairy products and other fatty foods where the presence of copper greatly reduces the oxidative stability. Wet ashing is suitable for most samples, though simple acid extraction from butter has been used with success. Several wavelengths are available, but the most sensitive is at 324·7 nm.

The analyst is interested in the iron content of foods, not because it is toxic, but because it is added to many foods as an essential mineral and plays a part in advertising. Because concentrations are high, the atomic absorption sensitivity is good, requiring only a simple extraction with hydrochloric acid. In the presence of citric acid, certain precautions must be observed owing to interference.

In sharp contrast, mercury is extremely toxic at low levels in most chemical forms. In recent years methyl mercury has been shown to accumulate selectively in fish. Wet-ashing methods are not suitable because typical concentrations are too low, although trace levels may be determined by a flameless method. Most total-mercury determinations are undertaken by extraction in acid with dithizone and chloroform, followed by a colorimetric method. Since the presence of organically bound mercury is of greater significance than that of elemental mercury, specific methods have been developed for the presence of methyl mercury in fish and marine oils, which involve gas chromatography as the iodide (or other halide) with electron capture for detection.

Lead is a traditional metallic poison, and most countries have prescribed legal limits for the lead content of foods. Apart from the natural content, most of the lead in foods can be traced to container contamination, particularly from soldered joints which have been attacked by acidic components in the food. Wet digestion with nitric acid/perchloric acid mixtures is suitable for sample preparation for flame analysis. Accurate determination by flame atomic absorption is limited to about 0·05 µg per ml, and therefore a concentration step is recommended. If a carbon rod atomizer is available, the detection limit is reduced to 0·2 ng per ml for a 5-µl sample.

Although tin is apparently not toxic, even at high levels in the diet, the main requirement for tin analysis in food samples is the control of corrosion in canned foods. At high levels tin may lead to taints in foods, and some countries specify legal limits in products. One recommended procedure for its analysis in foodstuffs involves hydrochloric acid hydrolysis and determination in a nitrous oxide-acetylene flame, which avoids interferences found in food samples.

From this outline of atomic absorption methods for the analysis of heavy metals, it should not be concluded that they represent the only procedures which are available. Thus, although chromium may be determined by the procedure already outlined, the detection and measurement of chromium in biological materials may also be undertaken by spectrophotometry, arc-emission spectrography, spark-source mass spectroscopy, colorimetry, polarography, neutron activation analysis, x-ray fluorescence or gas chromatography.

(e) *Vinyl chloride*
The carcinogenic nature of vinyl chloride has been recognized only fairly recently, as a result of the death of workers in factories which produce the plastic polyvinyl chloride (PVC). For many years, little attention was paid to the residues of vinyl chloride in PVC, and levels of more than 100 ppm were apparently quite common. Since the carcinogenicity of vinyl chloride was recognized, it was realized that these residues could find their way into foodstuffs packed in PVC wrappings, and efforts were rapidly made to reduce the level of the monomer in the plastic. This has been done very quickly by manufacturers in Britain without the need for government legislation.

While manufacturers were busy reducing the level of the monomer in PVC, analysts were quick to devise methods for its analysis. Since vinyl chloride is a gas, it lends itself very well to gas chromatography.

Analysis of naturally formed compounds in foods

(a) *Nitrosamines*
The discovery of the carcinogenic properties of N-nitrosamines by Magee

and Barnes in 1956 led to the development of analytical methods for their analysis. The formation of these compounds by the reaction of nitrous acid with secondary amines has been known since the dawn of modern organic chemistry, and it has formed the basis of a reaction for distinguishing primary, secondary and tertiary amines, a test which has been carried out by countless students.

$$CH_3-NH_2 \quad \text{primary amine}$$

$$\begin{array}{c} CH_3 \\ \diagdown \\ NH \\ \diagup \\ CH_3 \end{array} \quad \text{secondary amine}$$

$$\begin{array}{c} CH_3 \\ \diagdown \\ CH_3-N \\ \diagup \\ CH_3 \end{array} \quad \text{tertiary amine}$$

$$\begin{array}{c} CH_3 \quad H \\ \diagdown \diagup \\ N \\ \diagup \\ CH_3 \end{array} + HNO_2 \rightarrow \begin{array}{c} CH_3 \\ \diagdown \\ N-N=O \\ \diagup \\ CH_3 \end{array} + H_2O$$

dimethylamine   nitrous acid   dimethyl-N-nitrosamine

Formation of dimethyl-N-nitrosamine from a secondary amine

Under the conditions in which the reaction is performed, only secondary amines of the type shown in the scheme are claimed to produce nitrosamines, the essential structure of the amine being a single hydrogen atom attached to the nitrogen atom in the reactant. In practice it is found that tertiary amines will also yield nitrosamines in the presence of nitrous acid by loss of an alkyl group.

In the presence of small amounts of acid, nitrites yield nitrous acid, which is than capable of reacting with secondary amines from any environmental source. Such a system exists in traditionally cured foodstuffs, where meat or other protein foods containing many free amines are treated with sodium nitrite and sodium nitrate.

The problems associated with the analysis of N-nitrosamines in complex mixtures arise because there are relatively few highly specific chemical reactions of N-nitrosamines which are capable of distinguishing them from many other compounds. Methods of analysis are most valuable if they can detect individual nitrosamines when they are present in concentrations of a few parts in 10^9. Although in a cured-food product many secondary amines may be present (and the reactions by which nitrosamines are formed are certainly considerably more complex than the outline would suggest) in practice the N-nitrosamines which have been generally detected are restricted to a few simple compounds of the type shown.

$$\begin{array}{c} CH_3 \\ \diagdown \\ N-N=O \\ \diagup \\ CH_3 \end{array}$$

dimethyl-N-nitrosamine

$$\begin{array}{c} CH_2-CH_2 \\ | \quad\quad | \\ CH_2 \quad CH_2 \\ \diagdown \diagup \\ N \\ | \\ N=O \end{array}$$

N-nitrosopyrollidine

Nitrosamines of the type illustrated are volatile in steam, so that the bulk of food matrix may be separated effectively by this procedure. Further concentration may be effected by fractional distillation of aqueous solutions of nitrosamines, by prior addition of alcohol and collection of a narrow fraction boiling just above the temperature of the water-alcohol azeotrope. Solutions of nitrosamines in water are often unsuitable for chemical analysis, and it is usual to extract the aqueous solution with an immiscible organic solvent before further work is carried out. As dimethyl nitrosamine is completely miscible in both water and in a range of organic solvents, it is usual to add inorganic salts like potassium carbonate and sodium chloride to the aqueous phase in order to "salt-out" the nitrosamine into the organic solvent. Methylene chloride is the usual choice as the extracting solvent, since it is completely immiscible with water; it has a low boiling-point, so that extracts may be readily concentrated without undue loss of the nitrosamines, and it is the most efficient extractant for low-molecular-weight nitrosamines.

Concentration and purification of nitrosamines has also been carried out by adsorption on a column of carbon, but recoveries are variable according to the structure, and the process is far from selective. According to a recent procedure, which is currently used by several laboratories for the analysis of volatile nitrosamines in fried bacon, the food product is slurried in an aqueous salt solution, and the mixture is steam-distilled. The distillate, which contains the volatile nitrosamines, is saturated with salt and extracted with methylene chloride. The organic extract is then carefully evaporated; hexane is added towards the end of the evaporation, since the presence of methylene chloride is deleterious to the next stage of analysis.

N-nitrosamines exhibit two peaks in the ultraviolet spectrum, at around 350 nm and 230 nm, the latter being the more intense.

The determination of the concentration of nitrosamines in complex mixtures at very low levels presents problems which can only be solved by the use of sophisticated and expensive equipment. A laboratory contemplating this type of research needs to be equipped with a high-resolution mass spectrometer coupled to a gas chromatograph.

Although the application of this method provides a unique way of measuring N-nitrosamine concentrations at very low levels, it is unproductive to tie up £40 000-50 000 worth of equipment to carry out what is little more than routine analysis. Much effort has been spent in devising less-expensive equipment for this analysis. One approach has been the use of a gas-chromatographic detector which responds more or less specifically to nitrogen compounds. It has been known for several years that a flame ionization detector, modified with an alkali-metal halide (e.g. rubidium chloride) tip shows specificity to nitrogen-containing organic compounds. This principle has been developed to give a very stable nitrogen-specific detector, which relies on the formation of cyanide radicals in the flame for its operation. Although this type of detector shows some promise, natural extracts generally require further treatment for the removal of amines before analysis can be undertaken.

More recently there has been a further approach to the problem, which is based on completely different principles. The thermal energy analyser is sensitive to both volatile and non-volatile nitrosamines and can be specific to the N-nitroso group. The nitrosamine, dissolved in a suitable solvent like dichloromethane, is injected into a flash vaporizer, where the decomposition products are swept into a pyrolysis oven with argon gas. A specifically developed catalyst selectively breaks the N-NO bond, and the products enter a reaction chamber, held at reduced pressure, into which ozone is admitted. Under these conditions, an excited NO_2^* molecule is produced, which rapidly decays back to its ground state, with the emission of near infrared radiation which is readily detected and measured.

(b) *Mycotoxins*

(i) *Aflatoxins.* Mycotoxins are toxic substances produced by moulds which occur on various types of foods and feeds. The most widely studied group of mycotoxins are the aflatoxins, of which at least eight are known. The biological and chemical aspects of aflatoxins have been studied extensively, since they were shown to be the factors which caused Turkey X disease, which killed 100 000 turkeys in Britain in 1960.

Considerable importance is attached to the analysis of aflatoxins in peanuts; food companies rigorously reject samples which contain measurable levels of these toxic compounds. The peanuts are defatted in a Soxhlet apparatus by extraction with petroleum spirit, and a portion of the dry material is extracted with chloroform, aqueous acetone, or a mixture of chloroform and methanol. After purification (e.g. by thin-layer chromatography) in which an appropriate band is removed and extracted, the concentrated extract is further examined by thin-layer chromatography on plates of silica gel.

A procedure which is widely used for the quantitative examination of aflatoxin levels in extracts is known as *dilution to extinction*. Extracts are progressively diluted until the spot due to aflatoxin B (on kieselgel G, a type of silica gel) is just visible. The amount of aflatoxin present is then 0·4 nanogram.

For the analysis of aflatoxins in cottonseed products, the sample is blended with a mixture of acetone and water, a solvent that affords a quantitative extraction of aflatoxins, essentially free from lipid contamination. The crude extract is treated with lead acetate solution, filtered and extracted with dichloromethane, and chromatographed over alumina. The evaporated eluate is dissolved in a suitable solvent and submitted to thin-layer chromatography on silica gel with chloroform-acetone (9:1) as the developing solvent. Quantitative measurements on the spots are carried out by ultraviolet fluorescence, and compared with standards of aflatoxins B_1 and B_2 (G_1 and G_2 also, if thought to be present). The presence of the aflatoxins can be confirmed by spraying the plate with dilute sulphuric acid; under long-wave ultraviolet radiation, aflatoxins B_1 and B_2 exhibit a yellowish fluorescence, whereas G_1 and G_2 show a more blue-coloured fluorescence. The method is capable of detecting a few parts in 10^9 of the aflatoxins in cottonseed products.

In the fluorescence of a spot relied on for the determination of aflatoxins, false positives will sometimes occur. It is therefore desirable that confirmatory tests should be devised. Any spot which shows more intense fluorescence under short-wave than under long-wave ultraviolet radiation is not an aflatoxin. The use of dilute sulphuric acid has already been mentioned. Trifluoroacetic acid reacts quantitatively with aflatoxins B_1 and G_1, but B_2 and G_2 are unaffected.

High-pressure liquid chromatography has also been suggested as an analytical method for the examination of aflatoxins. Silica gel is generally used as the adsorbent, with mixtures of methanol and chlorinated hydrocarbons as the mobile phase; this type of system is capable of resolving the four main aflatoxins, but the sensitivity is probably insufficient to detect levels below 10 in 10^9, whereas good thin-layer chromatographic methods will detect 1-2 in 10^9.

Aflatoxins M_1 and M_2 generally accompany the other toxins, and their determination in milk is of particular importance, since they are formed as metabolites of the other toxins in contaminated feeds. Recently a method has been published for the detection of aflatoxin M in milk at the level of parts in 10^{12}. Milk, treated with cadmium sulphate solution, is extracted with chloroform, and the organic extract is chromatographed over a column of silica. The aflatoxin is eluted from the column and analysed by thin-layer chromatography, and the appropriate spot is observed under UV radiation.

(ii) *Other mycotoxins.* Patulin is a toxic metabolite produced by several species of *Aspergillus* and *Penicillium*, and it has antibiotic, carcinogenic and mutagenic properties. Many reports have shown that patulin occurs in apples contaminated by *Penicillium expansum* and also in apple juices. Rigorous methods for the analysis of patulin, particularly those applicable to apple juice, are clearly important.

Fortunately, patulin has a comparatively low molecular weight and can be submitted directly to analysis by gas chromatography. However, much better peaks are obtained if the patulin is analysed as a derivative, and for this purpose the acetate and trimethyl derivatives have been suggested. The patulin derivatives exhibit a good sensitivity towards electron capture; a single injection of a solution which contain 0·1 ng of patulin (as the derivative) provides a measurable response. This method has also been applied to extracts from mouldy rice.

However, the mere appearance of a peak on a gas chromatogram at a given retention time does not provide absolute proof of the identity of a compound. When the results of a series of analyses may influence future legislation, more substantial evidence is required; and, for this purpose, the analyst will naturally turn to the use of the mass spectrometer. The mass spectrum of patulin trimethylsilyl ether shows peaks at ions with mass to charge ratios (m/e) of 226, 211 and 183. The appearance of these ions in a mass spectrum of a scan taken at the appropriate moment during a gas chromatography-mass spectrometric run supplies the necessary evidence for its existence.

But the use of the mass spectrometer in this dynamic mode reduces the ultimate sensitivity of the instrument to nanogram quantities, and this may not be sufficient for the detection of patulin at the parts in 10^9 level in the product. A much greater sensitivity is achieved if the spectrometer is operated in the static mode, i.e. when it is focused at a specified m/e value. In the case of trimethylsilyl patulin, it is clear that, if the spectrometer is focused on any of the three quoted ions, it is unlikely that any compound other than patulin will produce a peak at the retention of patulin. Many modern mass spectrometers are constructed so that it is possible to switch rapidly among several chosen ions; this technique has been termed *multiple ion monitoring* or mass fragmentography. The presence of ions at m/e 226, 211, and 183, all appearing at the retention time of trimethylsilyl patulin, yields almost unassailable evidence for its presence in a mixture, particularly when the known relative intensities of the three ions are also taken into account. The method may also be used quantitatively by measuring the absolute heights of the peaks, followed by comparison with known concentrations of the authentic patulin derivative.

Zearalenone is an estrogenic secondary metabolite produced by a number of species of *Fusarium* colonizing maize and other cereals. It is of practical importance because feedstuffs contaminated with it cause hyperestrogenism when ingested by pigs.

Clearly there is also a possibility that contaminated cereals may enter the human food chain. In thin-layer chromatography, zearalenone can be located under short-wave ultraviolet radiation by its natural fluorescence, and its position on the plate slightly above the aflatoxin B_1 spot. The fluorescence of zearalenone is considerably less intense under long-wave UV radiation, and this feature can be used as an aid to identification.

(c) Polycyclic hydrocarbons

Polycyclic hydrocarbons are a group of compounds whose structure consists of a series of benzene rings fused together. They are produced in small amounts during the pyrolysis of almost any organic matter, and are thus very widely distributed in the air, soil, animal and plant kingdoms, geological sediments, and in a broad spectrum of foods, including smoked foods, roasted coffee, some roasted meats, and other commodities which come into contact with waxes and mineral oils. Members of this class of hydrocarbon have been shown to be carcinogenic to laboratory animals, and are suspected carcinogens for man.

The analysis of polycyclic hydrocarbons is complicated because of their very low concentration, and the variety of other organic compounds with which they are associated. Since the compounds are so widely distributed, and the levels in foodstuffs so low, it is important that only the purest solvents are used in the analysis, and that all the glassware is scrupulously clean.

A recent procedure, by which it is claimed that fourteen polycyclic hydrocarbons can be examined at low levels in meat, involves digestion (of meat and fish samples), followed by extraction with cyclohexane and separation by a succession of chromatographic techniques.

Each hydrocarbon possesses a strong absorbance in the UV part of the spectrum, and this property is then used for the final analysis. Each member of a group of hydrocarbons has a maximum absorbance measured in nanometres at a slightly different point in the spectrum, and the amount of absorbance is a direct measure of the concentration of the polycyclic hydrocarbon.

In order to determine the hydrocarbons in the fractions, the UV spectrum of each is recorded and compared with spectra of authentic compounds. Polycyclic hydrocarbons present in the original foodstuff at a level of 2 parts in 10^9 (2 µg per kg) are detectable. Excellent separations of polycyclic hydrocarbons have been achieved also by thin-layer chromatography on acetylated cellulose with mixtures of solvents such as methanol-ether-water. The positions of the compounds on the TLC plate are easily recognized since they fluoresce strongly under UV radiation.

Food additives

Food Dyes and Antioxidants
The analysis of dyes and antioxidants in foods is in a rather different category to those already discussed, since both are deliberate food additives which have been thoroughly tested and found quite safe.

Foods are analysed for their synthetic colour content generally on a qualitative basis to determine whether the individual dyes present comply with the laws of a particular country; it is worth noting that these vary considerably from country to country, so that a colour which is allowable in one may be considered illegal in another. It is unlikely that a gross excess of a colour in a foodstuff would be missed by visual inspection.

Antioxidants which are legally added to foods are fairly cheap, so it is unlikely that a manufacturer would substitute some relatively unknown chemical, but it may happen that a well-known antioxidant in one country is banned in a certain foodstuff in another country. Quantitative errors can, however, be made during the manufacturing stage, so that analyses also need to be quantitative.

Various procedures have been suggested for the analysis of food colours; most rely on paper and thin-layer chromatography. According to one procedure which is applicable to the detection of most food dyes, the food product is intimately mixed with Celite and packed into a chromatographic column. It is found that the dyes can be stripped from this column by elution with a liquid ion-exchange resin in an acidic organic solvent. The particular ion-exchange resin consists of a weakly basic high-molecular-weight secondary amine which combines with the food dyes to give products which are preferentially soluble in organic solvents. The eluate is freed from interfering impurities by addition of hexane, followed by washing with water. The dyes are then partitioned into aqueous ammonia, and the solution is concentrated under reduced pressure. The acidified extract is chromatographed over a column of polyamide with a mobile phase of ethanolic aqueous ammonia. Identification of the individual colours is achieved by paper chromatography, by means of an elimination process which involves the use of a selection of developing solvents, and the extensive application of known standards. In some instances paper chromatography is unable to distinguish between some colours; in these cases resort is made to thin-layer chromatography on microcrystalline cellulose or silica gel.

Conclusions

Rarely is there only one method of analysis for a particular component in a product; rather there are several available, and the preceding review should not be regarded as exhaustive. Part of the job of a good analyst is to decide which method is the most appropriate for his use.

Highly sophisticated equipment will often allow a particular test to be done more quickly. Against this, the analyst must weigh up the cost of using such expensive equipment, and the fact that the method may be an untried one. Some analyses, based on very simple measurements, are now known to give slightly incorrect results but, since plant managers and workers in a similar category have long accepted these results and their significance, confusion might result if more-accurate analytical methods were introduced.

Although the choice of an analytical method must in part be influenced by the available equipment, due consideration must also be given to the expected level of the component under analysis. Thus, it would be virtually impossible to carry out any analysis in the parts/10^9 region by ultraviolet spectroscopy. On the other hand, it would be wrong to use a mass spectrometer for the quantitative analysis of a substance in the percentage range.

The final choice must lie with the analyst in charge of the work, where his main function is to ensure that the results which are obtained are as accurate as possible within the requirements of the work.

FURTHER READING

General
Fishbein, L., *Chromatography of Environmental Hazards*
Vol. 1. *Carginogens, Mutagens and Teratogens* 1972
Vol. 2. *Metals, Gaseous and Industrial Pollutants*, 1973
Vol. 3. *Pesticides*, 1975
Elsevier Scientific Publishing Co.

Pesticides
Zweig, G., ed., *Analytical Methods for Pesticides and Plant Growth Regulators*, Academic Press, Vols. 1–8.

Nitrosamines
(a) Scanlan, R. A. (1975), "N-Nitrosamines in Foods," *C.R.C. Critical Reviews in Food Technology*, pp. 357–402.
(b) Crosby, N. T. and Sawyer, R. (1976), "N-Nitrosamines. A Review of Chemical and Biological Properties and their Estimation in Foodstuffs," *Advances in Food Research*, **22**, pp. 1–71.

Polychlorinated Biphenyls
Hutzinger, O., Safe, S., and Zitko, V. (1972), "Polychlorinated Biphenyls," *Analabs Research Notes*, **12** (2), 1–11.

Food Dyes
Endean, M. E. and Bielby, C. R. (1975), "The Extraction and Identification of Artificial Water-Soluble Dyes from Foods," B.F.M.I.R.A. Research Report No. 214.

Mycotoxins
Campbell, T. C. and Stoloff, L. (1974), "Implications of Mycotoxins for Human Health," *J. Agr. Food Chem.* **22** (6), 1006–1015.

Heavy Metals
Pinta, M., ed. (1975), *Atomic Absorption Spectrometers*, Adam Hilger.

CHAPTER THREE

SOME ASPECTS OF WATER QUALITY CONTROL

D. B. JAMES

Introduction

The environment in which modern man lives is the result of a complex process which began with the origins of life on earth and, but for the presence of water, could not have developed as we now know it. All living things require water to survive, from desert plants that germinate and flower briefly after infrequent rain, to man himself whose dependence on water to drink and to produce food is fundamental. When we consider man's needs within the environment he has created for himself, the problems of water quality assume alarming proportions.

Primitive man realized very early in his development that water was essential to his existence; he needed water to drink, the plants and animals he required for food flourished only where there was an adequate water supply. He found that water could be useful as a defence against his enemies, that it was a useful source of power, and that too much could lead to chaos and destruction. In modern times, the horrors of drought in Ethiopia and the results of severe flooding in Bangladesh are vivid illustrations of the important place of water in the environment.

For thousands of years man has been aware of his need of water in the right quantity but, as his technology developed, as urbanization began, so pollution commenced, and slowly the need was realized for water of adequate quality. It is a sad fact that in the industrialized societies of the world, greed and technology have outstripped science. This has resulted in many rivers and water courses becoming heavily polluted by urban and industrial effluents, and the population of towns and villages dependent on a river for water becoming exposed to epidemics of water-borne diseases and to excessive amounts of trace metals, because of the lack of knowledge and equipment to monitor and control the quality of water in the environment.

This lack of knowledge is best illustrated by the fact that the science of bacteriology was not fully appreciated until the mid-nineteenth century, and for several decades after this many water supplies continued to be drawn from rivers downstream of effluent discharges which received little or no treatment; annual rounds of typhoid and cholera were a regular feature of life and death. Control of water quality by careful selection of sources and treatment processes has virtually eradicated water-borne epidemics in the developed countries, but there still remains a problem in the third world.

The lack of equipment is shown in the current concern over lead in the environment. It is only in the last 20–30 years that reliable equipment has been readily available to measure accurately lead (and other elements) in very low quantities and, more important, to assess the effect of lead on the human body (see Vol. 6).

Today the human race appears to have reached the point at which it has realized many of its mistakes in the misuse of water, and is laying down standards, recommendations and guidelines, to control water quality at various stages in the water cycle in an attempt to reverse a trend which, in developed countries, had almost reached disaster level.

The water cycle

As can be seen from figure 3.1, the fraction of the water cycle we are concerned about here is the 3% existing as fresh water, and it is this 3% that is generally available to man for purposes of health, food (agriculture and freshwater fisheries), industry and leisure. Because of this, measuring, monitoring and controlling quality in this portion of the cycle is important to all societies.

Before considering how to do this in detail, it is important to consider what happens in nature to affect the quality of water (contamination) and what man has superimposed on this (pollution).

Contamination

It has been estimated that the upper portion of the earth's crust consists of 95% igneous rocks, made up of quartz and complex silicates such as felspar, and 5% sedimentary rocks, the most important of which are shales, sandstones and limestones. These rocks are subject to mechanical and chemical weathering, in which water plays an important part. Mechanical weathering can take place by abrasion when water flows over a hard surface, or when water in a fissure freezes and expands, thus creating

SOME ASPECTS OF WATER QUALITY CONTROL

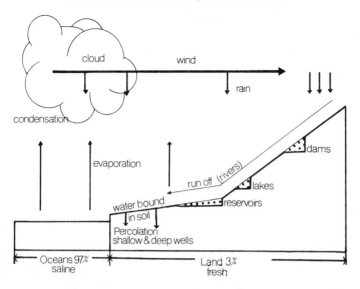

Figure 3.1 The water cycle

tremendous hydraulic pressures and cracking. Chemical weathering takes place by the leaching out of soluble material from porous rocks and soils. Hence natural waters contain naturally-derived material in solution and suspension.

Although many elements are present in natural water, the dissolved mineral matter consists mainly of the cations calcium, magnesium, sodium and potassium, and the anions bicarbonate, carbonate, sulphate, chloride and nitrate. In addition to these, oxides of iron, aluminium and silicon may be present in significant amounts. Organic materials derived from soil and decaying vegetation may also be present in quantities that affect the colour and taste of the water—a well-known example of this can be found in water flowing over Scottish heather moors which imparts a characteristic yellow-brown colour ot the water, highly valued for making and mixing with Scotch whisky.

Another source of dissolved constituents in natural water is rain water, which is a weak chemical solution containing traces of chloride, sulphate, nitrate, sodium, magnesium, potassium and dissolved gases such as oxygen, nitrogen, carbon dioxide, sulphur dioxide and hydrogen sulphide.

We can now begin to build a picture of natural water, not just as the H_2O of our early school science but H_2O plus a number of trace contaminants, the variety and concentration of which depend very much on the water's geological history.

Our natural water also contains a wide selection of living material—viruses, bacteria, algae, and small animals (both vertebrate and invertebrate), some of which may be pathogenic or parasitic.

Thus, depending where we draw our sample of natural unpolluted water, it will vary considerably in its quality, and may be anything from safe and palatable to dangerous. For example, fluke disease (bilharzia) affects man in Africa and in India, and the eggs of the causative organism are spread by contamination of water and soil. The embryos on hatching require an intermediate host, usually a snail, and the developed larvae give rise to a sporocyst. This multiplies and then releases large numbers of small white hair-like organisms which survive 48 hours in water, during which time they can affect humans if they gain access to the body by ingestion or by burrowing through the skin.

Pollution

In addition to natural constituents, the various uses made of water by man also add to the number of substances likely to be found in it.

Man's requirements for water vary, depending on the area in which he lives, from less than 50 litres per day in a primitive society to over 500 litres per day in a highly-developed industrialized society. The water balance of the average human adult is shown in Table 3.1.

Table 3.1 The human water balance

Intake	Output
1650 ml drink	1700 ml urine
750 ml food	500 ml perspiration
350 ml produced by the body from food	400 ml breathing
	150 ml faeces

The remaining quantities of water are used for agriculture, industry, and domestic and personal hygiene.

Thus a modern city of one million population could require as much as 500 megalitres per day of potable water to meet its needs, and as much as 90% of this would have to be dealt with as effluent. Domestic effluents contain a variety of substances such as body wastes, detergents, oils and fats, depending on the degree of development of the area. Industrial wastes may contain a vast array of materials—fertilizers, herbicides, insecticides, fungicides, acids, oils, and other synthetic chemicals.

Nearly all of these effluents have to be discharged to a water course (some are recycled by industry) and, unless they are monitored and controlled, chaos would result, as often one man's effluent is another man's drinking water. Figure 3.2 illustrates a typical river story from source to ocean, and illustrates how a river can become an open sewer if water quality is not efficiently controlled at the effluent level, and how huge populations would be at risk without correct water treatment.

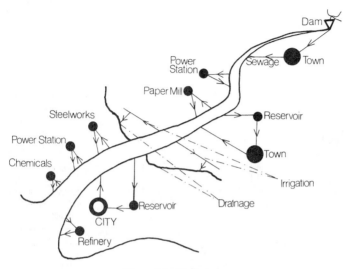

Figure 3.2 Typical river

The above considers only man's physical needs; modern society is constantly pressing for more freshwater-based leisure and recreation—sailing, canoeing, fishing, swimming, etc. None of these can be pleasantly or safely pursued unless we control the quality of the aquatic environment.

Measurements necessary for control

In order to control the quality in any part of the water cycle, it is first necessary to measure the concentrations of contaminants and/or pollutants, to assess the effect of these on man and his environment, and then to decide how best to control them at acceptable levels. Table 3.2 illustrates a range of determinations that may be required, together with the discipline and method of analysis that may be employed. It is not possible to give a complete list here as the number of determinations required to cover all cases is vast, but it serves to illustrate the range that has to be considered.

Before describing techniques it is perhaps profitable to look at the units of concentration used in the field of freshwater analysis. The concentration units most used in expressing the results of chemical analysis are shown in Table 3.3.

Table 3.2 Typical list of parameters and their significance in water analysis.

Parameters	Significance	Discipline	Method of Analysis
colour	A	1	colorimetric
turbidity	A	1	photoelectric
taste	A	1	dilution
odour	A	1	dilution
temperature	A	1	mercury or electric thermometer
pH	B	1	electronic
electrical conductivity	B	1	electronic
dissolved solids	B	1	gravimetric
suspended solids	B+D	1	gravimetric
residual disinfectant	B+E	1	colorimetric
total hardness	B	1	titrimetric
calcium	B	1	titrimetric or A.A.
magnesium	B	1	titrimetric or A.A.
sodium	B	1	A.A. or emission spectroscope
potassium	B	1	A.A. or emission spectroscope
aluminium	B	1	A.A. or colorimetric
alkalinity	B	1	titrimetric
sulphates	B	1	gravimetric, titrimetric
chlorides	B	1	titrimetric
fluoride	B	1	specific electrode or colorimetric
nitrate	C+D	1	colorimetric
nitrite	C+D	1	colorimetric
ammonia	C+D	1	colorimetric
organic nitrogen	C+D	1	colorimetric
silica	B+D	1	colorimetric
material soluble in chloroform	B	1	gravimetric
silver	C	1	A.A.
arsenic	C	1	A.A.
barium	C	1	A.A.
cadmium	C	1	A.A.
chromium	C	1	A.A.
copper	C	1	A.A.
iron	C	1	A.A.
mercury	C	1	A.A.
manganese	C	1	A.A.
nickel	C	1	A.A.
lead	C	1	A.A.
zinc	C	1	A.A.
cyanide	C	1	colorimetric

SOME ASPECTS OF WATER QUALITY CONTROL 43

Table 3.2   continued

Parameters	Significance	Discipline	Method of Analysis
phosphates	B+D	1	colorimetric
hydrogen sulphide	C	1	colorimetric
mineral oils	C	1	G.L.C.
aromatic hydrocarbons	C	1	G.L.C.
phenols	C	1	colorimetric
detergents	C	1	colorimetric
pesticides/herbicides	C	1	G.L.C.
dissolved oxygen	B,D+E	1	specific elctrode or titrimetric
biochemical oxygen demand	C,D+E	1	
total organic carbon	C	1	combustion
total coliforms	E	2	multiple tube or membrane filtration
Escherichia coli	E	2	
faecal streptococci	E	2	
Clostridium welchii	E	2	multiple tube
Salmonella	E	2	multiple tube
colonies growing on agar	E	2	direct plating
viruses	E	2	concentration + plating
algal count	D	3	microscopic
aquatic animals	D	3	microscopic
biotic index	D	3	microscopic + macroscopic
radioactivity	C	1	Geiger or scintillation counter

INDEX:

    1. chemistry
    2. microbiology
    3. biology

    A. organoleptic
    B. physico-chemical
    C. undesirable or toxic
    D. biological
    E. microbiological

    A.A. atomic absorption
    G.L.C. gas liquid chromatography

One litre is the volume of 1 kilogram of water at 4°C at a pressure of 760 mm of mercury. Thus the density of water is 1·0 unit of mass per unit of volume in the metric system. Hence milligrams per litre are commonly referred to as the parts per million (ppm) or parts/10^6.

Occasionally the unit milliequivalents per litre may be used when the analyst wishes to show the balance of cations and anions in solution. One

Table 3.3 Units used in chemical analysis

Unit of concentration	Abbreviation	Relative concentration
milligrams per litre	mg/l	1 in 10^6
micrograms per litre	µg/l	1 in 10^9
nanograms per litre	ng/l	1 in 10^{12}

Table 3.4 Milliequivalents per litre compared to mg per litre

Anions	symbol	mg/l	meq/l
nitrogen	N	1	0·0715
chloride	Cl	1	0·0282
sulphate	SO_4	1	0·0208
Cations			
calcium	Ca	1	0·050
magnesium	Mg	1	0·0412
sodium	Na	1	0·0435

Table 3.5 Units used in microbiological analysis

Unit	Type of examination for which used
organisms per millilitre	bacteria on nutrient agar, algae
organisms per 100 millilitres	coliforms, streptococci
organisms per litre	*Salmonella*
plaque forming units/litre	viruses

milliequivalent of a cation is associated with 1 milliequivalent of anion. Examples of this system are shown in Table 3.4.

It is also important to define how a particular constituent is being expressed. Thus it is normal to express nitrate (NO_3), nitrite (NO_2), ammonia (NH_3) and organic nitrogen all in the same unit, i.e. nitrogen mg/l. Hardness salts (calcium and magnesium) are usually expressed as mg/l equivalent to calcium carbonate; and most trace elements as the element, e.g. iron (Fe) as mg/l Fe.

In biological and microbiological analyses the units normally encountered are shown in Table 3.5.

Ranges of concentrations found in various waters can be seen in Table 3.6.

It is impossible to describe each determination in the space available, but examples given below illustrate some commonly-used techniques.

SOME ASPECTS OF WATER QUALITY CONTROL

Table 3.6 Typical water analyses at source

	Irish impounding reservoir	Yorkshire lake	River Thames	Chalk well
colour (Hazen units)	17	112	20	0
pH	5·6	6·9	7·9	7·0
ammonia (N) mg/l	0·033	0·04	0·31	0·00
organic nitrogen mg/l	0·00	0·13	0·26	0·00
nitrate (N) mg/l	0·00	0·3	4·6	3·6
permanganate value mg/l	2·4	3·4	3·0	0·10
total solids mg/l	40	81	480	378
total hardness mg/l	8	46	294	318
alkalinity mg/l	2	35	203	270
chloride mg/l	10	7	29	18
sulphate mg/l	0	8	65	30
magnesium mg/l	2	6	5	40
iron mg/l	0·15	0·28	0·1	0·02
manganese mg/l	0·00	0·04	0·03	0·00
phosphate mg/l	0·001	0·005	2·1	0·50
coliform per 100 ml	2	365	14 000	8
E. coliforms per 100 ml	0	360	5288	4
colony count per ml at 22°C	7	114		45
colony count per ml at 37°C	2	26	5620	3

Chemical analyses

Titrimetric or volumetric analysis. This consists of determining the volume of a solution of accurately known concentration (the standard solution) which is required to react quantitatively with a solution containing the substance to be measured. The mass per volume of the substance to be determined is then calculated from known laws of chemical equivalence.

The standard solution is placed in a graduated vessel (usually a burette) and added to a known volume of the unknown solution until a visible change occurs in the mixture (normally a colour change produced by the presence of an indicator dye).

Total hardness is determined in this fashion and relies on the fact that a solution of EDTA (ethene diamine tetra-acetic acid) reacts with calcium and magnesium salts at a pH of 10. The completion of reaction can be followed by the presence of Eriochrome dye.

A solution of EDTA is made such that 1 ml is equivalent to 1 mg of calcium carbonate. A known volume of sample is placed in a glass flask, the pH adjusted, Eriochrome added, and a red wine-coloured solution is produced. EDTA is added until the red colour suddenly changes to a distinct blue. The volume of EDTA is noted and the following calculation completed:

$$\frac{1000}{v} \times \text{ml EDTA} = \text{mg/l CaCO}_3$$

where v is the volume of the sample in millilitres.

Thus if a 50-ml sample is used and 14·5 ml EDTA are required, the hardness would be 290 mg/l $CaCO_3$.

Colorimetry. Colorimetric analysis is a most valuable analytical technique; it relies on the application of the Beer-Lambert law which states that

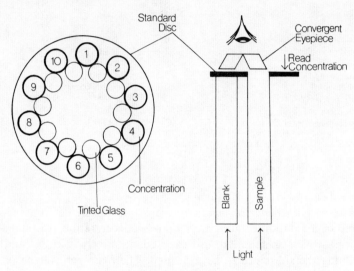

Figure 3.3 Nessleriser

the reduction in intensity of light passing through a coloured solution is proportional to the concentration of the colouring agent and the depth of the solution or

$$\frac{I_0}{I_t} = Kct$$

where I_0 is initial intensity, I_t final intensity, c concentration, t depth, and K a constant.

This principle is put to use by comparing under well-defined conditions the colour produced from the substance being analysed with the colour from the same substance in known concentration.

The intensity of colour produced can be measured by eye, comparing the unknown with a series of standards, or use can be made of a simple colorimeter, e.g. the Lovibond Nessleriser coupled with a series of artificial standards (figure 3.3). Use can also be made of the photoelectric cell, which produces an electrical output proportional to the incident light, to measure light intensity, and in an instrument designed for this purpose light of a preselected wavelength (colour) is used. The colour is selected by means of a filter or a monochromator, such as a diffraction grating. Figure 3.4 illustrates this type of equipment, which may vary from a very simple filter

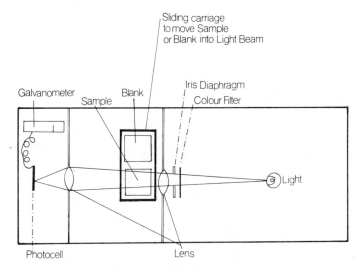

Figure 3.4 Simple photometer

photometer to a sophisticated spectrophotometer. When using this type of instrument, it is normal to calibrate the equipment for a particular determination by measuring light intensities with a series of standards, noting the meter readings (absorbance), and plotting a graph of these against concentration.

Figure 3.5 shows a calibration graph used for the determination of aluminium by the reaction of catechol violet dye with aluminium in solution at a pH between 6·0 and 6·2. This determination can also be carried out using a Nessleriser and standard disc.

Turbidity can also be measured by the photoelectric cell but, as can be seen in figure 3.6, instead of measuring transmitted light, this instrument relies on detecting light scattered by the particles causing turbidity. The degree of light scatter from the sample is measured against a series of turbidity standards.

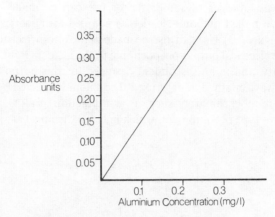

Figure 3.5 Aluminium calibration using a simple photometer and Ilford 607 filter.

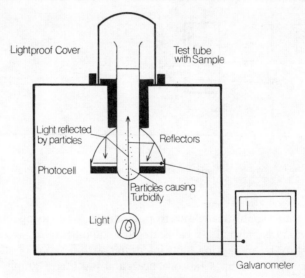

Figure 3.6 Turbidity meter

Emission spectroscopy. When a solution or solid is subjected to a high temperature, the elements in the sample become excited and emit light of a specific wavelength; the intensity of light at a particular wavelength is proportional to the concentration of the element producing that wavelength.

This is well illustrated in the determination of sodium by the use of flame photometry. A solution of sodium sprayed into an oxygen-acetylene flame causes the flame to emit an intense yellow colour which can be measured by a photoelectric cell. This colour is related to concentration by means of a calibration graph made by spraying a solution containing known sodium levels into the apparatus.

Gravimetric analysis. This analytical technique is a means of determining the concentration of a substance by direct weight.

Thus, if we need to know the amount of sulphate in a particular sample of water, this can be done by adding a solution of barium sulphate to a known volume of the sample at a temperature just below boiling-point. A white precipitate of barium sulphate is formed, which can be filtered on to paper, placed in a weighed crucible, heated to red-heat to burn the paper and re-weighed. Thus the mass of barium sulphate is found and, as the sulphate is 41·2% of this mass, the sulphate in the sample can be determined.

Specific electrodes. By placing specially-designed electrodes in a sample, it is possible to measure the concentration of a particular constituent due to the electrical voltage generated by it acting on the electrodes. The voltage produced is proportional to the concentration of the substance in the solution and thus, by coupling the electrodes to a sensitive millivoltmeter, the unknown can be found.

One of the most important analyses carried out in water quality testing is pH. The pH scale is from 0 to 14, a value of 7 being neutral, 0–7 acid, 7–14 alkaline; 0 is the most acid condition, 14 the most alkaline.

The degree of acidity or alkalinity is due to the presence of the hydroxonium ion (H_3O^+) in a solution which is produced by the reaction between water and a hydrogen ion from an acid.

The hydroxonium ion reacts with a special glass electrode to produce a small voltage which is measured by reference to a standard electrode immersed in the same solution. The result is shown on a scale graduated in pH units from 0 to 14.

Odour and taste. These two factors are most important in water analysis. They are determined by establishing the threshold taste or odour number by diluting the sample with taste-and-odour-free water (produced by passing water through activated carbon) until the nose or tongue cannot detect any difference between diluted sample and the taste-and-odour-free water. Thus, if a sample of water was thought to have an odour, it would be diluted with varying quantities of taste-and-odour-free water until the odour disappeared.

Radioactivity. Radioactivity is the result of energy released when changes occur at atomic level and is measured in curies. 1 curie (Ci) is equal to $3·7 \times 10^{10}$ disintegrations per second (becquerels (Bq)).

There is a natural background level of radiation harmless to man but, as a consequence of the rapid growth of atomic energy in the past 20 years, a new range of potentially hazardous substances has entered the environment.

In water, levels of radioactivity are generally very low (less than 10^{-8} microcuries/ml) but it is often necessary to determine the radioactivity of water if a potential hazard exists. When such an analysis is necessary, a large volume (1–10 litres) is evaporated to dryness and examined for alpha radiation by a scintillation counter, and for beta radiation by a Geiger-Müller counter.

Microbiological examination of water

It is rare to find a natural water free from bacteria, as they gain access to all points of the water cycle. Rain collects airborne bacteria; water passing over the earth's surface collects bacteria from the soil, vegetation and animal life.

For the general purpose of water examination we are interested in organisms which are mainly of excremental origin. These are (*a*) the coliform group, (*b*) faecal streptococci, (*c*) anaerobic spore formers. It is also of interest to determine the total number of bacteria in a given volume of sample.

It must be stressed that the routine examination for bacteria is concerned with the detection of bacteria which are generally harmless, but which are normal inhabitants of the human and animal intestine. They are so abundant in mammalian dejecta that pollution of water by extremely small quantities can be easily demonstrated. Also they greatly outnumber pathogenic bacteria, and can be used as indicator organisms as, if they are present, then dangerous organisms may also be present; if absent, it may be generally assumed that the pathogens are absent.

Bacteriological examination is the most sensitive test that is used to detect pollution of water by sewage.

The coliform organism. The most important determination made is that of the coliform bacteria, in particular *Escherichia coli* which is by far the most abundant organism found in faecal matter and sewage. Its detection in a 100-ml sample of drinking water should immediately raise suspicions of pollution likely to endanger health. It is found at levels between 100 000 to 1 000 000 per ml in sewage, and 1000 million per gram in human faeces.

Other types of coliform are also common in the intestines of man, mammals and birds, but they are also found in decaying vegetation. They have a sanitary significance, but are by no means as important as *E. coli*.

On a routine basis the coliform is detected by either a multiple tube or a membrane technique.

Multiple-tube-test. In this technique different amounts of water are mixed with liquid medium, incubated for 48 hours at 37°C, and then examined for signs of growth. The media used for detecting the organisms may be McConkey broth, Teepol broth or Improved Glutamate media, the last being considered best at the present time in the United Kingdom. Each tube of medium contains a small inverted tube of medium, and the medium itself contains a dye sensitive to pH changes. After 48 hours, coliforms are presumed to be present in tubes which have produced gas (bubble in inverted tube) and acid (colour change in dye). Small portions of the positive tubes are then inoculated into further tubes of medium (lactose ricinoleate broth) and incubated for a further 24 hours at 37°C and 44°C, growth at the former temperature showing the presence of coliforms, and the latter confirming the presence of *E. coli*.

Membrane filtration. In this technique, a known volume of water (usually 100 ml of a public supply, less from a suspect source) is filtered through a fine cellulose disc. The disc is then placed on a pad soaked in a suitable medium (Membrane Teepol broth) in a small covered dish. Duplicate discs and pads are prepared; one is incubated at 37°C, the other at 44°C for 24 hours, after which time the discs are examined. All yellow colonies are counted, and a small quantity picked off the surface with a Nichrome wire loop and placed in a tube of medium (lactose ricinoleate) for confirmation, as in the multiple-tube techniques.

Confirmed coliforms at 37°C are expressed as colonies per 100 ml and similarly confirmed *E. coli* at 44°C.

As indicated in Table 3.2, the last two techniques can be used to determine a variety of organisms utilizing different media.

Bacteria growing in agar. This test is used to show bacterial levels associated with a particular source of water. It does not indicate total bacteria.

Two separate 1-ml quantities of water are mixed with 15 ml of molten yeast extract agar at 45°C in a flat dish. The contents of the dish are allowed to cool, and one dish is incubated at 22°C for 72 hours, the other at 37°C for 24 hours. All colonies of bacteria growing are counted and the result expressed as colonies per ml in y days at x°C.

Viruses in water. The subject of viruses in water is under intense investigation at the present time. They are present in smaller numbers than bacteria in polluted water, and are identified by concentration and inoculation into tissue cultures together with antibiotics. Incubation may be longer than 14 days, and the number of viruses recorded as plaque-forming units per x litres.

Virus examination is extremely specialized and must be carried out in suitably equipped laboratories.

Biological examination

The biological assessment of water consists of the microscopic examination of samples for the presence of minute plants and animals, and the macroscopic examination of bodies of water for the presence of higher plants and animals. By examining a water sample, the biologist is able to establish a pattern of flora and fauna in a river or lake which can be useful in assessing the degree of pollution, or the effects of known pollution. He can establish biological patterns along a river near an intake to a water supply or an effluent discharge, and follow changes taking place in natural or man-made lakes.

Several techniques have been developed for the biological inspection of waters (two examples are given).

(a) *The Utermöhl method*—The equipment used is shown in figure 3.7. A sample of water of between 1 and 10 ml is placed in the tube, and 1 ml of Lugol's iodine solution added. The sample is allowed to stand for 1–10 hours, the upper portion of the apparatus removed, a cover slip placed over the concentrate, and the deposited organisms are examined through the base using an inverted microscope. The microscope eyepiece is fitted with a graticule, as the field is too large to count all organisms in one area. The number of organisms per ml and the variety of organisms are then recorded.

(b) *Membrane filter method*—A 1–20 ml sample is treated as above with Lugol's iodine and then filtered through a membrane filter, 25 mm in diameter. The membrane is placed on a microscope slide on which 1 drop of cedar wood oil has been placed, and dried in the dark at 55°C. After this operation the membrane will be clear; a cover slip with a drop of oil underneath is placed over it, and the specimen examined by a conventional microscope. The membrane is divided into quadrants for ease of counting; organisms/ml and type are noted.

Significance of water analysis

As indicated in Table 3.2, the parameters important in water analysis can be divided into five categories.

A. *Organoleptic factors*

These are the qualities which affect the aesthetic properties of a water: appearance, palatability and temperature. They are of paramount importance to a public water supply and the recreational use of water.

SOME ASPECTS OF WATER QUALITY CONTROL

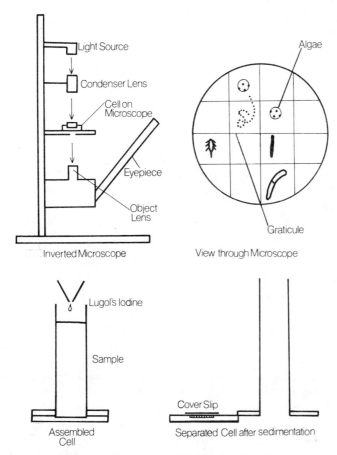

Figure 3.7 Utermöhl technique for counting algae.

B. *Physico-chemical factors*

Substances which are not generally a danger to health (but are important because above set levels they may cause a nuisance or have economic consequences) fall into this category, e.g.

pH
A low or high-pH water can cause corrosion of concrete or metallic objects which are in contact with it.

The pH can also influence the type and quantity of flora and fauna living in water, and thus affect the suitability of a water for treatment to public supply standard and its recreational value.

Hardness
The type and quantity of hardness salts (calcium and magnesium) in a water supply are of considerable economic significance.

The harder the water, the more soap is required to produce a lather, and the more likely is a hot-water system to be blocked by excessive scale formation.

Dissolved solids and conductivity
Both of these factors indicate how much residue would be left by evaporating a sample of water to dryness.

The higher the dissolved solids the more brackish a water becomes. This is undesirable for domestic and industrial use as it can affect palatability, suitability for irrigation, and the operating cost of steam-generating equipment.

Aluminium
Aluminium compounds are widely used in water treatment (see page 64). If the residual aluminium becomes excessive, sediments form in the water.

C. Undesirable or toxic factors

Substances in this group are those which may be a danger to the health of plants and animals.

Many substances fall into this category if they are present in large amounts, but in the aquatic environment concern is for substances which are of significance when present in microgram quantities.

In this group are the heavy metals, which are known to be a danger to most forms of life, and the many chemicals produced by modern technology. Petroleum products, pesticides, herbicides, detergents and a host of other manufactured products offer a hazard to the water cycle.

The majority of substances in this classification gain access to the water cycle in effluent discharges from industry, domestic sewers and farm drainage. It is essential that their levels are monitored and controlled at all points at which they can gain access to the water cycle, for they could destroy the aquatic environment and present a danger to public health.

D. Biological factors

In this group are found the wide varieties of plants and animals present in aquatic ecosystems, and the substances which affect the growth or decay of a particular species within the ecosystem.

It is true to say that the less polluted a water is, the wider the variety of plants and animals that will be found. Thus it is important that not only are the numbers and types of aquatic plants and animals determined but also those substances which influence them.

Nitrogen (in the form of nitrite, nitrate, ammonia and organic matter), phosphates, traces of iron, other minerals and dissolved oxygen can increase the productivity of a river or lake. Conversely the uncontrolled discharge of effluents can reduce the oxygen level, introduce traces of toxic material into a watercourse, and reduce the variety of aquatic life to the point at which it becomes septic and completely unacceptable for most purposes.

Analysis for biological factors associated with public water supplies is concerned with monitoring events in large volumes of stored water in order to predict any difficulties in treatment which may reduce the quality of the water, e.g. taste due to decaying plants. The efficiency of a treatment process can be judged by biological examination.

Most public water-distribution systems contain small aquatic animals such as asellus, cyclops, shrimps, midge larvae and worms. As they feed on deposited organic matter, the extent to which they colonize water pipes is an indication of treatment efficiency and the internal condition of the pipes.

Another important aspect of biological examination is the investigation of parasites. Obviously parasites affecting human beings are most undesirable in a public water supply, and they also interfere with the recreational use of water. Those which are parasitic to animals reared or used for food are also important.

E. *Microbiological factors*

Bacteria play an important role in the aquatic environment, and a well-balanced population in a lake or watercourse has a vital role in converting many substances into nutrients useful to plants and animals.

Freshwater microbiology is mainly concerned with looking for bacteria and viruses indicative of pollution, and the bacteria and protozoa essential to biological treatment processes. Of paramount importance are organisms which can cause disease to humans (pathogenic organisms). The main diseases which may be water-borne are listed in Table 3.7.

Hence microbiological analysis of water is essential to public health, as the presence of pathogens renders a water unsuitable for drinking and for many recreational purposes.

Scope of water analysis

The purpose for which a water analysis is carried out may vary considerably, from assessing the suitability of a water for irrigation to its fitness for human consumption. In order to judge the suitability of a treated water (such as the river Rhine in Holland) for consumption, it may be necessary to carry out all the analyses in Table 3.2, and many others. On the other hand, if a public supply is gathered from an upland gathering ground, where it is

Table 3.7   Water-borne diseases found in polluted waters

Organism	Disease
Salmonella typhi	typhoid
Salmonella paratyphi	paratyphoid
Salmonella sp.	food poisoning
Vibrio cholerae	cholera
Shigella sp.	dysentery
Entamoeba histolytica	dysentery
Schistosoma sp.	bilharzia
Taenia sp.	tape worm
Leptospira sp.	Weil's disease
Enteric viruses:	
(a) Polio virus	poliomyelitis
(b) Coxsackie virus	myocarditis, Bornholm disease
(c) Echo virus	diarrhoea respiratory infections

known that there is no pollution, a simple examination for colour, pH, hardness, alkalinity, total solids, chlorine and coliforms may be all that is necessary.

Thus, before analysis it is essential to state the purpose of the examination.

Public water supplies

Of paramount importance in the freshwater cycle is the quality of water to be distributed, after suitable treatment, for human consumption. In the British Isles, water suitable for domestic supply is described as *wholesome*. This is defined as water taken from a properly protected source, submitted to an adequate system of purification so that it is free from visible suspended matter, colour, odour and taste, and from all objectionable bacteria indicative of the presence of disease-producing organisms; it contains no dissolved matter of mineral or organic origin which would render it dangerous to health.

Table 3.8 indicates standards applicable to public supplies by the World Health Organization. These are always under review and may vary from one publication to the next. The European Economic Community is currently drawing up a list of standards to be observed by member states. Standards alone cannot cover all eventualities, and therefore it is of vital importance that a constant watch is kept on the gathering ground, so that any activity which is likely to affect water quality is monitored and controlled, e.g. siting of new industry, and control of effluent quality.

SOME ASPECTS OF WATER QUALITY CONTROL

Table 3.8 World Health Organization drinking water standards (1971)

Substance	Highest desirable level	Maximum permissible level
colour	5	50
total solids	500 mg/l	1500 mg/l
turbidity units	5	25
pH	7·0–8·5	6·5–9·2
total hardness	100 mg/l	500 mg/l
chloride	200 mg/l	600 mg/l
phenols	0·001 mg/l	0·002 mg/l
sulphate	200 mg/l	400 mg/l
copper	0·05 mg/l	1·5 mg/l
iron	0·1 mg/l	1·0 mg/l
manganese	0·05 mg/l	0·5 mg/l
zinc	5·0 mg/l	15 mg/l
nitrate (N)		10·2 mg/l
fluorides	0·60–1·7 depending on temperature	
arsenic		0·05 mg/l
cadmium		0·01 mg/l
cyanide		0·05 mg/l
lead		0·1 mg/l
mercury		0·001 mg/l
selenium		0·01 mg/l
polynuclear aromatic hydrocarbons		0·2 µg/l
coliforms	0	0 (3*)
E. coliforms	0	0

* in non-disinfected supplies.

Recreational use of water

In industrial and urban communities, outdoor recreation is a preferred form of leisure for millions of people, and the availability of water in respect of quantity, quality and accessibility plays a vital role in satisfying this demand. People like to use water to walk by, sit near, fish in, swim in and boat on, and in order to pursue these pastimes the quality of water available must meet certain standards.

In considering the quality of water suitable for recreation, a number of factors have to be taken into account.

In general terms, the water should appear to be aesthetically acceptable, of reasonable clarity, free from turbidity that could settle and cause objectionable deposits, free from floating debris; scum, foams and oil should be absent, and there should be no objectionable odour. Substances which could cause undesirable physiological response in animals, fish or plants, e.g. radioactivity or herbicides, should be absent, as should substances that can produce undesirable aquatic life, e.g. high nitrates and phosphates which encourage massive weed and algal growth.

Bacterial levels are also important; a recent American Government Committee recommended that, for recreational purposes involving swimming, diving and water skiing, the average *E. coli* level should not exceed 200 per 100 ml, and that a single maximum figure of 400 per 100 ml should be the highest allowable level. For other recreational purposes, such as walking, boating and camping, a mean of less than 1000 *E. coli* per 100 ml was considered satisfactory.

Thus for recreational use, water of a fairly high standard is required, with special attention being paid to bacteria, dissolved and suspended solids, temperature (too warm upsets the biological quality), pH (prevention of eye irritation) and clarity (safety of swimmers).

Aquatic life. A tremendous variety of plants and animals live in or near water. Most of these are able to adapt to slow changes in water quality but, now that man has the ability to alter his environment drastically, the aquatic environment may change more rapidly than the plants and animals can adjust.

Substances in solution or suspension, whether solid, liquid or gas, determine the quality of water, and aquatic organisms can be affected directly or indirectly due to the competitor/predator situation which exists in their environment.

Assessing water quality in respect of aquatic life is extremely complex. Areas that require specific attention are dissolved oxygen, salinity, pH, dissolved nutrients and toxicity; it is also important that acceptable levels are maintained during low and high flow, and high and low temperatures, as any material may become toxic at a given level.

Because of the complexity of assessing the relationship between water quality and aquatic life, use is often made of a biotic index system in which water quality is graded against the type of life found.

One very generalized system was used by the UK Department of the Environment during the 1970 survey of river pollution. This is illustrated in Table 3.9.

Water quality in relationship to aquatic life is also important to recreation, as many leisure pursuits such as fishing and wildfowl shooting involve harvesting food. Man is involved in the aquatic food chain and, the better the aquatic environment, the safer the food.

Water is used for irrigating crops, watering livestock and (in modern society) the washing of fruits and vegetables in preparation for market. Particularly critical is the use of water in the production of milk. In areas remote from public water supply, the farm will have to provide itself with a supply suitable for all domestic purposes. Therefore the whole range of parameters in Table 3.2 may have to be considered to assess a water's suitability for agricultural use.

Table 3.9 Scheme of river classification (UK Department of the Environment 1970)

Class	Biota found in river—
A	Rivers with widely diverse invertebrate fauna, including caddisfly larvae, mayfly and stonefly nymphs, and freshwater shrimps. Salmon, trout and grayling if ecologically favoured, or a good coarse fishing with a variety of species.
B	Stonefly and mayfly nymphs not common. Caddisfly and shrimp present. Varied invertebrate population. Good mixed coarse fisheries, trout may be present but not dominant.
C	Rivers in which invertebrate population is restricted, dominated by the water louse. Some shrimps, caddis and mayfly rare. Fisheries moderate to poor, restricted to roach and gudgeon.
D	Invertebrate fauna absent or restricted to worms and chironimids (bloodworms). Incapable of supporting fish life.

Industrial use of water. The availability of water is critical to industrial development. Textiles, paper, chemicals, petroleum, coal, metal industries, engineering and the food industry, all require vast amounts of water. It was estimated in the United States in 1964 that approximately 49×10^{12} gallons per year were used by industry. 90 per cent was used for cooling or condensing purposes, 8 per cent in processing, and 2 per cent for boiler water feed.

The range of water quality suitable for industrial use varies from brackish waters of more than 1000 mg/litre dissolved solids (which can be used for cooling) to a public water supply used for food processing, and further treatment for boiler water feed.

Because of the variety of uses for water in industry, each industrial user, having secured the quantity of water required, usually treats the water using the full scale of modern technology to obtain the desired quality.

Monitoring of water quality

In monitoring water quality it is necessary to sample, analyse, report and assess the information so obtained. The resources available to monitor water quality are limited and the following points must therefore be considered when planning a monitoring programme.

(i) The variables of the system with which we are concerned must be laid down. In routine examination of water this is essential, as a vital parameter may be missed, or time wasted measuring a useless one. A study should be made of the catchment area of a body of water, so that natural and manmade impurities that affect quality can be assessed. For example, if a watercourse is to be used for irrigation purposes, the degree of salinity has to be assessed in relation to the crop grown, and any pesticides that may enter the watercourse and have a detrimental effect on plant growth must be measured.

In the routine analysis of water, laboratory resources are generally limited, and efficient use must be made of them.

(ii) The container and method to be used for sampling must be carefully considered. Of prime importance is a clean container of adequate volume. If a sample is to be taken to assess taste and odour, a glass container, cleaned with chromic acid and rinsed with odour-free water, is essential.

For bacteriological sampling a sterilized bottle must be used.

If the metal content of a water is required, it is preferable to add acid to a sample taken into a plastic bottle, so that no part of the determinand is adsorbed onto the surface of the container.

(iii) Very careful consideration must be given to the choice of sampling point. If the effect of an effluent discharge on a river quality is to be considered, the river must be sampled above and below the discharge, allowing for mixing to occur.

(iv) In order to obtain useful information, the correct frequency of monitoring is important. There are three basic types of monitoring, "grab" sampling, automatic sampling and analysis, and aerial surveyance.

"Grab" sampling consists of a sample taken manually, transported to a laboratory and analysed. It has the advantage that many analyses can be carried out on one sample, but is labour-intensive, and important changes can be missed between "grabs". Frequency can be increased by automatic sampling at pre-set intervals, e.g. hourly, but substantial laboratory backup is required to analyse samples so produced. Frequency, and the number of stations sampled, are proportional to cost.

In automated systems of sampling and analysis, a self-contained unit housed either on the bankside or submerged, continuously samples, analyses, and either stores or transmits information collected. This system has the obvious advantage of continuous monitoring, but suffers from the disadvantage that the number of parameters measured is limited, e.g. dissolved solids, temperature, suspended solids, pH and dissolved oxygen. The number of stations is proportional to the cost. Automatic systems must be supplemented by grab sampling.

Aerial cover to monitor pollution is extremely expensive and measures very few parameters, e.g. temperature by infrared photography and colour.

But it gives very wide coverage and may be very useful in specific circumstances.

In the 1972 survey of the Colorado river, the following costings were estimated:

Grab sampling at 22 stations giving 10% cover, $26 880 p.a.; automatic sampling at 22 stations giving 43% cover, $130 460 plus $2000 per station for recorders, or plus $6000 for a telemetry and logging system plus a terminal computer; aerial monitoring for 10% cover, $66 000.

(v) The amount of attention given to points (i)–(iv) is irrelevant if the laboratory analysing samples is inadequately equipped or staffed, as samples require rapid accurate analysis, expert interpretation, and inilation of any required action. Society must be prepared to invest people and money in order to improve the aquatic environment.

(vi) The method of analysis to be used should be carefully chosen, so that the desired balance of speed and accuracy is achieved. A rapid method of analysis is of no advantage if it cannot detect accurately at the critical level.

The importance of the above points cannot be over-emphasized when considering the establishment of a laboratory to measure and monitor any system.

Control of water quality

The measuring and monitoring of water quality is carried out by a wide variety of organizations. Those concerned with the environment are involved in assessing water quality from source to estuary in the short and long term, in detecting improvement and deterioration, in deciding the policies that will prevent good-quality watercourses deteriorating, and in improving the standard in those that have been neglected or unused.

It is important that, throughout the water cycle, quality is maintained at each point. Water possesses remarkable powers of self-recovery, but each user of water must play a part in maintaining water quality, so that rivers and lakes are not irreversibly damaged; subsequent users must receive a material suitable for their purpose.

Rain water

Control of water quality is necessary from the moment evaporation from sea or land occurs, as the atmosphere contains traces of sulphur dioxide and nitrogen compounds due to pollution from industrial conurbations. Rain precipitating in such an atmosphere may be very acid and slowly affect the area on which it falls. Atmospheric pollution can have far-reaching consequences; it may travel many thousand of miles in jet-stream currents and affect water quality in non-industrial areas. Swedish scientists are concerned about the long-term effect on moisture entering the atmosphere in the heavy-industrial areas of Europe, precipitating and reducing pH in Scandinavian lakes.

In the Pennine range of hills in Britain it was not unusual in 1963 to record rain-water pH of below 3, due to the presence of sulphuric acid originating from fossil fuels used in the Birmingham and Manchester conurbations. In more recent years the application of the Clean Air Acts has achieved a considerable improvement, and the water in the Pennine area is slightly less acid today than 15 years ago.

Atmospheric pollution may also involve toxic metals and harmful organic compounds; their control at source is an essential part of water quality control.

Lakes

Water impounded in either natural or man-made lakes has to be managed carefully, or changes may occur to the detriment of quality, particularly if discharges containing phosphates and nitrogen compounds are made into or upstream of the lake.

When an accumulation of pollutants occurs, the dissolved oxygen level falls, and natural thermal stratification takes place, so the activity of anaerobic bacteria in the bottom muds increases, releasing a range of nutrients into solution. An increase in nutrients means increased plant growth, which eventually decays and contributes to the sediments. The cycle is repeated year after year, and slowly eutrophication takes place, accompanied by a massive increase in algal activity. If this is linked with bacterial pollution from domestic sewage, the environment around the lake becomes both unpleasant and unhealthy.

Most of London's water supply is derived from the RiverThames above its tidal reaches. Although effluents discharged into the river are well treated, the water abstracted from the river and stored in large artificial lakes contains many nutrients. To prevent the growth of undesirable algae, and to assist the treatment process, a careful watch is maintained by the biologists, chemists and microbiologists employed by the Thames Authority. When the reservoirs are beginning to stratify, and dissolved oxygen falls to an unacceptable level, the water is artificially mixed by water jets coupled to pumps; thus problems are minimized.

The Great Lakes of North America have suffered very badly from pollution in the last hundred years. In Lake Michigan, which may be taken as a typical example, discharge of domestic and industrial waste from hundreds of communities has spoilt the recreational value of the area and made some beaches a health hazard.

Efforts are being made to reverse the situation; Chicago's sewage is now discharged into a watercourse flowing into the Mississippi river, and a programme of research is under way to determine means of improving the Lake water quality; but it is a slow and costly process.

The Lake of Zurich is an important recreation area for almost a million people; it is also an important water source, and receives the sewage effluents from several towns. The population around this area has increased by 240% since 1850, and the lake has become steadily more and more eutrophic. A reduction in detergent phosphates and extensions to sewage works are preventing rapid deterioration, but a constant watch is being kept by the responsible bodies.

Rivers

The maintenance and improvement of the quality of water in rivers has become increasingly important in recent years due to social and economic pressures.

In terms of cash and lost land, the cost of building dams has escalated alarmingly in the last decade and, in order to meet the increasing demand for public water supplies, it has been realized that if the quality of a river is improved, it can be directly utilized as a source, or it can be used to transport water from dams in upstream catchment areas to the point of use, thus saving money in pipeline construction.

The flow of water in the River Severn in Britain is regulated by controlling the flow from large dams in its catchment area. By controlling the effluent discharges from both domestic and industrial water users, the water quality in the river is maintained at a level which allows many communities to use it as a water source.

At the turn of the century the Sacramento river in North America became badly silted due to hydraulic mining. The silt ruined salmon spawning beds, but with the cessation of mining the river slowly improved and salmon again returned. In recent years, due to the importance of forestry in an area upstream from the spawning grounds, a paper mill was constructed, but such is the quality of the effluent from the mill that the salmon are left unaffected. This illustrates that with thought, wildlife and industry can co-exist.

In 1957 a survey of the London reaches of the Thames showed that, excluding eels, there were no fish at all in this stretch of the river. Due to the improvement in river quality, brought about by the application of legislation relating to effluent discharge, 80 different species of fish have been observed in the past 10 years.

Underground water

In addition to surface water, quality in underground sources has to be carefully monitored, as it can also be affected by events on the surface.

Water in natural subterranean reservoirs may have travelled many miles and percolated through many different materials in rural and urban locations.

There has been concern in recent years because of increasing nitrate levels in some underground waters, particularly in those areas where, due to intensive farming, nitrates are leached from the soil into underground waters. This can also happen in areas where there are natural nitrate deposits within the catchment area, e.g. parts of Israel and Ethiopia.

The dumping of industrial wastes can also have an undesirable effect on water quality. Until 1972 several companies discharged effluents into old mineworkings in Lanarkshire (Scotland). Eventually the effluent raised the water level, and it discharged through a disused minemouth on the banks of a nearby river causing gross pollution. This incident led to the passing of the Clyde River Purification Act 1972 which allows the River Authority to control discharges to underground strata.

The provision of a safe water supply

For whatever purpose it is to be used, water quality is important, but of paramount concern is the provision of a safe water supply for domestic use.

The provision of a safe supply involves processing a raw material of variable quality to a final product which must be consistently of the highest quality. Purification of water involves a number of operations to improve the physical, chemical, bacteriological, virological and biological properties of a source to an acceptable level of hygiene.

Each operation can be considered a barrier, and one or more barriers may be required.

First, the source of water must be *protected* from as much pollution as possible. In practice this usually means employing national or federal legislation applying to all water users.

Secondly, the water should be *taken into a reservoir system*, allowing variations in quality to average out and settlement of suspended deposits to occur, giving an enormous reduction in bacterial and organic components.

The third is *coagulation*, which consists of adding aluminium or iron salts to the water, and adjusting the pH to a level at which they become insoluble. As the precipitate forms, fine suspended material, bacteria, and some organic and inorganic material in solution are enmeshed in or adsorbed by the newly formed particles, which can then be removed by sedimentation.

In some cases sedimentation is not required, but a barrier found in most treatment processes is *filtration*. Filtration following sedimentation or coagulation is normally carried out by passing water through a fine sand bed to remove all matter in suspension. The water passes through a rapid filter at a velocity between 0·8 and 1·6 mm/s.

Another form of filtration used when a water requires a large improvement in bacteriological and biological factors is slow filtration through sand at a velocity of approximately 0·03 mm/s.

In most cases, before a water is suitable for domestic use, it is necessary to *disinfect* it. This may involve the use of chlorine, ozone, ultraviolet light or chlorine dioxide in order to ensure a water free from pathogenic bacteria and viruses.

If water is drawn from a source which contains industrial pollutants of potential danger, or compounds giving objectionable taste or odour, it may be necessary to pass it through a filter containing *activated carbon*, which is able to remove a wide variety of objectionable substances by adsorption.

Finally, in particular cases, *special treatments* may be given. If the water is too hard for domestic purposes, it may be softened or, should it contain too many dissolved solids, these can be reduced by ion exchange processes.

A water purification process must be designed to suit the requirements of a particular case. Only by the careful selection of barriers can a safe supply of water be guaranteed.

FURTHER READING

Holden, W. S., ed. (1970), *Water Treatment Examination*, J. & A. Churchill, London.
   A comprehensive textbook covering all aspects of water supply from source to tap; very readable, but the reader should have a scientific background.
Analysis of Raw, Potable and Waste Waters, HMSO, 1972.
   This is a detailed publication of methods used in chemical analysis.
Wallwork, J. F. (1972), "Waterworks Laboratory Equipment," *Journal of the Society of Water Treatment and Examination*, **22**, 17–40.
   This article indicates the range of equipment required in a water laboratory.
The Bacteriological Examination of Water Supplies, HMSO, 1969.
   An excellent publication giving a background to water bacteriology as well as detailed methodology.
International Standards for Drinking Water, World Health Organization, Geneva, 1971.
   As well as details of standards, this publication gives a general background to water quality criteria.
Deeming, H. G. (1975), *Water. The Fountain of Opportunity*, Oxford University Press, New York.
   The nature and uses of water are presented. No significant aspect of water's environmental influence is omitted. Very readable, with minimum technical knowledge required.

CHAPTER FOUR
HAIR AS A MIRROR OF THE ENVIRONMENT

JOHN LENIHAN

HAIR IS A TISSUE OF INTEREST AND VALUE TO STUDENTS OF THE ENVIRONMENT FOR SEVERAL REASONS:

1. It records contamination from inside the body (since it is a route of excretion for many metals) and from outside, since it traps metallic vapours and dust.
2. It acts as an integrating dosemeter over a period of months; human head hair grows at about a centimetre per month. Blood and urine, widely studied as indicators of exposure to toxic substances, indicate only recent exposure.
3. Many heavy metals are found at relatively high concentrations in hair, because of their affinity for the proteins of which hair is largely composed.
4. Extremely sensitive techniques are now available for the estimation of metals in hair.
5. Hair samples are easily collected, stored and transported without deterioration—in marked contrast to most other tissues.
6. Hair is very durable; samples up to 1900 years old have been analysed without difficulty, giving useful information about the internal chemical environment in earlier times.

To understand how these properties come about, it is necessary to know something of the physical and chemical structure of hair.

Growth and structure of hair

Hair growth begins in the hair follicles which cover the body. There are roughly 100 000 follicles on the scalp, between 10 000 and 20 000 on the bearded area of the face, and between 500 000 and 1 000 000 on the rest of the body. In the follicles, which lie just below the skin, hair is formed—like other tissues—from materials in the blood and other body fluids. As growth proceeds, the hair is extruded from the follicle as a fibre, and rises above the skin. The visible hair is not a living tissue; increase in length results from the continual extrusion of freshly formed material from the follicle.

A typical hair consists of a *root*, embedded in the follicle, and a *shaft*. The shaft is made up of the *cuticle* (a thin outer layer of scales), the *cortex* (a hollow cylinder of keratin—a name applied to a group of proteins) and the *medulla*, a central cavity. The medulla is filled, to a greater or less extent, by cells of irregular shape and by tiny air bubbles. In old age the cells shrivel, and the air bubbles become more numerous.

The hair then reflects light more effectively; for this reason—and because of loss of pigment from the cortex—old people usually have white hair. The diameter of a typical hair is about 0·1 millimetre, and the mass of a 1-centimetre length of a single hair is about 100 micrograms.

The process of keratinization occurs also in skin and nail, two tissues which are in many ways similar to hair. Living cells formed in the hair follicle, the lower layers of the skin or the nail bed, are gradually converted into keratin. Keratinous tissues are therefore composed essentially of dead cells, but have not lost all of their vital activities. Many substances, including mercury and iodine, can be passed into the body by rubbing with ointments; other substances can penetrate the outer layers (as vapour or liquid) and become bound to keratin.

Head hair and chest hair grow at a little over 1 centimetre a month. Beard hair and pubic hair grow roughly half as fast. About 90% of the hairs on the human head at any time are in a growth phase, the remainder being in a resting phase, preparatory to falling out. The growth phase lasts about 900 days and the resting phase about 100 days, so that the average life of a head hair (if not cut or combed out) is about 1000 days. A typical head contains about 100 000 hairs; the average daily loss is therefore about 100 hairs. A life of 1000 days corresponds to a length of about 40 centimetres; since hair which is not cut will grow longer than this, some hairs presumably undergo more than one growth cycle before falling out. A typical man of Caucasian race grows about 35 milligrams of beard hair per day; in Japanese the rate is only 10–12 milligrams per day. Finger nail grows at about 0·1 millimetre per day, and toe nail at less than half of this rate.

Metals in hair

The value of hair as a mirror of the chemical environment depends on the fact that it is composed largely of keratin, which is a mixture of proteins. Many metals can be readily attached to protein molecules. Metal binding is important in the operation of enzymes, many of which need metal ions to fulfil their catalytic functions. The toxicity of many metals is related to their ability to react with proteins and, for example, to displace (or inhibit by becoming attached to) essential metals from critical sites on enzyme molecules. The binding of metals to hair protein does no harm, but gives useful information about the medium from which the hair was formed, i.e. the circulating blood. Hair can also be contaminated by sweat, by impurities in air and water, and by soaps, dyes and other cosmetic preparations.

Hair accumulates trace metals to a greater extent than many body tissues. One factor in this process is the abundance of cystine, an amino acid which makes up about 14% of human hair. The cystine molecule is double-ended, with two sulphur atoms (the disulphide bridge) in between. Proteins consist of chains of amino-acid molecules. There are 26 amino acids, which can be arranged in any order; for this reason the number of different proteins is very large.

In keratin, some of the many cystine molecules are shared between adjacent amino-acid chains, having one end in each chain, linked by the disulphide bridge. These cross-linkages are largely responsible for the strength and elasticity of hair. In one method of permanent waving, some of the disulphide bridges are broken by treatment with a reducing agent, which converts the —S—S— structure into two SH-groups. The hair then loses some of its elasticity, allowing remodelling by heat, after which treatment with an oxidizing agent removes hydrogen from the SH-groups and allows disulphide bridges to reform, giving a measure of permanence to the new style.

Many metals are found in hair, bound to sulphur atoms in cystine and to sulphydryl (SH) groups present in other amino acids. Table 4.1 shows typical concentrations in hair, blood and whole body. It will be seen that, for many metals, the concentration in hair is higher than the average concentration throughout the body.

Table 4.1 Trace element concentrations in whole body, blood and hair

	Typical concentration, ppm		
	Whole body	*Blood*	*Hair*
Al	0·9	0·14	5
As	0·1	0·004	0·2
Br	3	4	30
Cd	5	·01	1
Co	0·03	0·0004	0·1
Cr	0·1	0·03	1
Hg	0·2	0·01	2
Mn	0·3	0·01	0·3
Ni	0·1	0·04	3
Pb	1·5	0·2	20
Sb	0·04	0·1	0·2
Se	0·2	0·1	1
Ti	0·2	0·03	4

Analysis of hair

There are a number of analytical techniques suitable for the estimation of trace elements in hair at or below the level of 1 ppm ($1/10^6$). Since a typical hair sample can seldom exceed a few milligrams in mass, the analyst is looking for very small fractions of a microgram of the elements under investigation. Consequently much care is needed to avoid contamination in gathering the samples and in handling procedures before analysis.

A single hair is not representative. For many trace elements the concentration varies by a factor of two or more among different hairs from the same head. This variation is not surprising. A hair in the growing phase will contain trace elements recently abstracted from the blood; hair in the resting phase will not. Consequently different hairs reflect metabolic activity from different periods of time. The trace-metal content of hair is influenced also by atmospheric contamination and by cleansing, and other materials applied to the hair. Except on a sparsely covered head, this contamination is not likely to be uniformly distributed.

Hair removed by combing is likely to be in the resting phase. A more representative sample is obtained by cutting. Distribution of a trace element is not uniform along a single hair. Analysis of successive portions of a hair reveals peaks and troughs corresponding to fluctuations in dietary intake or external contamination during the period of growth; this technique is sometimes useful in forensic or historical studies (see figure 4.1 on page 85).

Efforts to establish hair as a means of identification (by comparing trace-element concentrations in hairs found at the scene of a crime and in hairs taken from a suspect) have been successful only to a limited extent. The trace-element content of an individual's hair can change appreciably (by a factor of two for some elements) over a period of a few weeks. Even if the suspect is found soon after the crime, the variation in trace element among hairs from the same head, and the fact that usually only one or two hairs are found at the scene of the crime, make positive identification difficult. Abnormal concentrations of trace elements, from cosmetics or industrial materials, can sometimes be used as an aid to identification.

In the study of poisoning, whether homicidal or occupational, the problem is quite different, since the trace-element levels in hair are usually many times greater than normal, and therefore outside the range of natural variability. In epidemiological work on large populations (e.g. to study the relation of environmental factors to the prevalence of particular diseases) the effects of individual variation can be allowed for by using population samples of adequate size.

In general it is satisfactory to gather a sample consisting of a bundle of hairs about the length and thickness of a matchstick—though much smaller samples can be analysed quite accurately if the need arises. The sample is best taken from the middle portion of the hair, so as to avoid excessive contamination from the scalp. If contamination by the scissors is likely to be significant (e.g. when analysing for iron, cobalt or nickel) the ends of the hairs may be trimmed with plastic scissors or with a sapphire knife.

The traditional methods of analysis, involving wet chemistry, are not suitable for the estimation of trace elements in hair. This is because conventional techniques are designed for the analysis of solutions; hair is not easily brought into solution, and the reagents needed to dissolve it are inevitably contaminated with trace elements in amounts which, though of no significance for the reagents themselves, may be comparable with the amounts of the same elements present in the sample under analysis. The analysis of hair is, therefore, best attempted by methods which require no preparatory chemical treatment.

It is, nevertheless, sometimes necessary to remove external contamination, loosely attached to the surface of the hair. For this purpose a number of treatments have been devised, involving quick rinsing in ethyl alcohol, acetone or ether. Organic solvents are preferable to detergents, which usually remove material from the body of the hair (as well as from the surface) and may themselves contain trace elements, such as arsenic and manganese, in appreciable amounts.

Virtually all of the elements contributing to environmental pollution can be estimated by one of two methods: activation analysis and atomic absorption. Both methods are essentially techniques for counting atoms, independent of their state of chemical composition; for the estimation of compounds, the analyst uses other methods, such as chromatography.

Activation analysis

Activation analysis is one of the few analytical methods which depend on the properties of the nucleus. Most of the chemical properties and reactions of materials are related to the behaviour of the electrons in the outer parts of the atoms. The nucleus is an assemblage of particles held together by strong forces—in contrast to the electrons, which are rather easily dislodged or rearranged.

In activation analysis we use the property that most nuclei are altered by bombardment with subatomic particles or radiation. The alteration takes the form of temporary radioactivity (i.e. a structural rearrangement, from which the nucleus escapes by emission of particles and/or radiation to form a new stable nucleus) not always of the same element as the original target.

The particles most often used for activation analysis are thermal neutrons, obtainable in large numbers at low cost inside a nuclear reactor.

Neutrons are released from uranium nuclei in the process of fission, the source of nuclear power. Thermal neutrons are so called because, having been repeatedly slowed down by collisions, their average kinetic energy is the same as that of the atoms and molecules of the medium in which they are moving; this energy depends on the temperature.

The activation process can be illustrated by considering a typical reaction:

$$^{23}_{11}\text{Na} + ^{1}_{0}\text{n} \rightarrow ^{24}_{11}\text{Na} \rightarrow ^{24}_{12}\text{Mg} + ^{0}_{-1}\text{e} + \gamma$$

The nucleus of a stable sodium atom has a mass number of 23 and an atomic number of 11, corresponding to a positive charge of 11 units—resulting, of course, from the presence of 11 protons. If a bombarding neutron is captured (as a small proportion are), the charge of the target nucleus is unaltered, but its mass is increased by one unit to make ^{24}Na. This isotope is unstable (or radioactive), and its nucleus decays by the emission of beta particles (electrons) and gamma rays to yield the stable isotope ^{24}Mg. The energies of the beta particles and the gamma rays are characteristic of the nuclear species producing them, and may be used for identification. The decay of a radioactive nucleus is exponential and is characterized by the half-life, i.e. the time required for the transformation of half of the nuclei originally present. The half-life of ^{24}Na is approximately 15 hours.

Many elements of environmental significance, including copper, manganese, mercury, cadmium and arsenic are readily activated by neutron bombardment; a few, notably lead, cannot be estimated conveniently by activation analysis.

If the neutron flux and the properties of the target nuclei are accurately known, it should be possible, from the measurement of the induced radioactivity, to calculate the mass of the target substance present in the irradiated sample—but this is only a theoretical possibility, not capable of realistic fulfilment.

In practice the quantitative analysis is achieved more easily. The sample (usually containing many elements which can be activated) is put inside the reactor, along with a separately wrapped sample of the element to be estimated—or a convenient compound. The induced radioactivity corresponding to the element of interest in the experimental sample is eventually compared with the activity of the reference standard irradiated at the same time; the mass of the desired element is then found by simple proportion.

The process of activation analysis is not quite as simple as it looks. A sample of hair (or of any other tissue) contains many elements which become radioactive under neutron bombardment. Some of the elements most readily activated (such as sodium and chlorine) are present in amounts hundreds of times greater than the trace element which it is usually desired to estimate. Consequently the irradiated sample displays a confusing pattern of radioactivity, with superimposed contributions from many elements and very small signals from the element of interest.

The required activity can be separated from the heavy background of unwanted radiation in two ways—instrumentally or chemically. The instrumental approach is by gamma-ray spectroscopy. Just as the constituent colours in a beam of light can be separated by optical spectroscopy, so a complex gamma-ray spectrum can be resolved into its components, each corresponding to a particular element. Unfortunately the resolution obtainable in gamma-ray spectroscopy is not as good as in optical spectroscopy. Since each activated element in an irradiated sample produces several gamma-ray lines, there are limits to the accuracy with which the spectrum can be analysed. In practice, gamma-ray spectroscopy is adequate for many elements at concentrations of a few parts per million, but is not often sufficiently sensitive to detect constituents present at concentrations below 1 part per million.

The alternative method is to separate the element of interest by chemical manipulation of the irradiated sample. It may be thought that this procedure defeats the purpose of activation analysis, which is to avoid contamination by reagents. However, the irradiated sample cannot be contaminated in this way, because the atoms of interest have been set apart by being made radioactive. The reagents do not contain radioactive isotopes and can therefore be used with impunity.

Indeed, the chemical separation goes a stage further. Only a small proportion (usually less than one in a million) of the atoms in a particular element are actually made radioactive in a typical activation procedure. This minute amount of material is liable to be adsorbed on the walls of the test-tubes and flasks in which the chemical separation is attempted. For this reason it is usual to add to the irradiated sample, as soon as it has been dissolved, a known amount (usually a few milligrams) of a carrier substance in the form of a compound of the element under investigation. Adsorption on the walls of the reaction vessels is then provided mainly by the carrier. The presence of the carrier has another advantage. At the end of the separation process (which is done by the traditional methods of analytical chemistry) the amount of carrier recovered can be weighed. It will usually be less than the original amount, because of the inevitable losses in the separation process, and a correction can be made for the yield.

The product of the separation process, containing most of the original carrier and the same proportion of the desired radioactive isotope resulting from the irradiation in the reactor, is then assayed for radioactivity. Since only a single isotope is now present, relatively simple Geiger or scintillation counting equipment may be used, in contrast to the elaborate equipment, including electronic computers, necessary for gamma-ray spectroscopy. The yield correction already calculated is applied to the measured radioactivity. The reference standard is put through the same separation process as the experimental sample with which it was irradiated. Eventually two counting rates are established; the ratio between them is the ratio between the mass of the appropriate element in the reference standard and the mass of the same element in the experimental sample.

Activation analysis is generally a very sensitive technique, but is not suitable for the estimation of every element. When thermal neutrons are used, the sensitivity is not usually adequate for the estimation of elements with atomic numbers below 10 or above 80. Activation analysis is well suited to solid samples, since no preliminary treatment is normally needed, but the irradiation of liquid samples in a nuclear reactor often presents difficulties because of the inevitable temperature rise.

Atomic absorption

Among the other analytical techniques suitable for the study of trace elements in biological materials, atomic absorption is probably the most useful. This technique depends, like every analytical method, on provoking the sample to display behaviour characteristic of the elements to be detected. Like many techniques, it depends on processes leading to the rearrangement of electrons, in contrast to activation analysis, which depends on rearrangement of the nucleus.

Atomic absorption belongs to the large group of optical techniques based on the study of the characteristic pattern of emission or absorption of light by atoms or molecules. A simple example is the flame test used by beginners in analytical chemistry. Atoms which absorb energy from the flame become excited, i.e. some of their electrons are moved from the ground state to higher energy levels. When the electrons drop back to the ground state, they emit the surplus energy as light quanta. The frequency of these quanta is characteristic of the atomic species involved, and it is therefore possible to identify the constituents of the sample by examining the emitted light. This can be done crudely by the naked eye, as in the school laboratory, or by a spectrometer, in which case about thirty elements can be detected. Flame emission techniques were used qualitatively as long ago as the 1860s and were made quantitative in the 1930s. They have limited usefulness, because only a very small proportion (usually no more than one in a million) of the atoms in the flame are excited at any given time. Another difficulty arises because the intensity of the emitted light is not proportional to the concentration of the element concerned in the flame. Some of the light emitted in the hotter central part of the flame is absorbed by atoms in the cooler outer regions before escaping. Consequently flame emission techniques need careful calibration.

Atoms in the ground state will absorb radiation at the same frequencies that they would emit if excited. Techniques depending on absorption are therefore potentially much more sensitive than those depending on emission, since every atom is a potential absorber, whereas only the very small proportion of atoms that have been excited are potential emitters.

Molecular absorption in solution is a technique often used for the estimation of compounds. The method here is to produce a suitable coloured compound which may be concentrated by extraction in organic solvents, and then to study the absorption of visible or ultraviolet light of appropriate frequency in passing through the solution.

Atomic absorption, like activation analysis, is a technique for the detection of atoms rather than molecules. The basic observation on which the technique depends was made by Wollaston, an English physician, as long ago as 1802, when he noticed that the continuous spectrum of light from the sun is crossed by a number of dark lines. In 1859 Kirchhoff explained the origin of the dark lines. Light emitted by excited atoms in the hot central part of the sun is absorbed by cooler atoms of the same elements in the outer layers.

For the practical application of atomic absorption as an analytical technique we require three things:

(a) a source of atoms,
(b) a source of light,
(c) a detecting system to study the absorption of light in passing through the sample.

Since atomic absorption depends on the absorption of light by free atoms (not molecules) it can be exploited only in the gas phase. A convenient way of obtaining a source of free atoms is to spray a solution containing the experimental sample into a flame. The solution quickly evaporates to give a dry salt which then vaporizes; some of the vaporized compound dissociates into atoms. A temperature between 2000 and 3000 K is usually suitable for these processes to occur.

A small proportion of the atoms in the flame will, of course, be excited and will emit light on their own account, but the great majority will remain in the ground state, ready to absorb light of the appropriate wavelength.

The most useful light source is the hollow-cathode lamp, a discharge tube containing argon or neon at a pressure of about 1 centimetre of mercury. The cathode is a hollow cylinder made of (or lined with) the element of interest or some suitable compound. The lamp usually needs an electrical supply of 400–500 volts and draws a current of 15–20 milliamperes. It is customary to use a separate lamp for each element.

A few experimental refinements, which will not be gone into here, are necessary to avoid confusion in the detecting system from light emitted directly from the flame. The equipment for atomic absorption has now reached a state of considerable sophistication, allowing accurate estimations to be made at a relatively rapid rate and without the need for a highly skilled operator.

Sensitivity of atomic absorption can be further increased by using a flameless source of atoms. Here the sample is quickly vaporized by electrical heating in a horizontal graphite tube, with a hole in each end through which the light beam passes. The improvement in sensitivity arises for two reasons. Firstly, the sample is vaporized in a very short time, whereas in flame atomic absorption the sample is introduced into the flame gradually. Secondly, the graphite cylinder used in flameless atomic absorption traps the free atoms, allowing a relatively large absorption signal to be obtained. In flame atomic absorption, the atoms are not confined, and therefore spend only a relatively short time in the optical path where they can contribute to the absorption signal. For many elements, flameless atomic absorption is 100–1000 times more sensitive than the flame technique.

Atomic absorption is suitable for the estimation of most metals, notably magnesium, calcium, zinc, cadmium and sodium, where its sensitivity approaches that of activation analysis. For other elements, the sensitivity of atomic absorption is less than that of activation analysis, but is often more than adequate. For a few elements, including arsenic, selenium and antimony, atomic absorption is not very sensitive; activation analysis is therefore the method of choice. For some elements, including lead, lithium and beryllium, activation analysis is not at all suitable, but atomic absorption is quite sensitive.

In reviewing the usefulness of these techniques for the study of hair, we shall consider first some problems of the contemporary environment, both internal and external, and shall then show how the analysis of hair illuminates the environments of times past.

Arsenic and smoking

The tobacco smoker makes a distinctive environment of his own, particularly in the lungs, where conditions are favourable for the uptake of trace elements into the circulation. In view of the generally accepted relationship between smoking and lung cancer, it is worth while to consider the relevance of carcinogenic substances in tobacco smoke. Cigarettes contain measurable amounts of arsenic, derived from the soil and from insecticides used to spray the tobacco plant. In 1957 some cigarettes were found to contain more than 50 ppm of arsenic, though a typical level today would be about 1 ppm. About 60% of the arsenic in a cigarette is retained in the ash, 25% remains in the unsmoked stump, and about 15% appears in the smoke. Tracer experiments with radioactive arsenic show a retention in the lungs of about 5% of the total arsenic in a cigarette. About 50% of the retained arsenic is excreted in urine and faeces within ten days; of the remainder, some may be expected to stay in the body, and some to be incorporated in the hair, which is a significant route of excretion.

It is not difficult to show that a smoker inhales considerably more arsenic than a non-smoker. The average amount of arsenic inhaled in a day in Britain (based on measured levels in air) is about 2 micrograms. From the figures given above, it would be seen that a person who smokes 20 cigarettes a day, containing 20 micrograms of arsenic, will inhale about a further microgram of arsenic per day.

A large-scale experiment was made in Glasgow in 1958 to find whether there was any difference in the arsenic content of hair between smokers and non-smokers. Samples from more than 1000 people were analysed, but no difference was found between smokers and non-smokers. It should, of course, be remembered that the average daily dietary intake of arsenic may be as much as 100 micrograms.

Additional evidence for the retention in the lungs of inhaled arsenic (whether from the atmosphere or from tobacco smoke) was obtained in another experiment conducted in Glasgow in which five mice spent a normal life-span in a dust-free atmosphere, and another group of five inhaled the air of Glasgow, containing about 3×10^{-12} gram of arsenic per litre. The diet and living conditions were the same for both groups. The average arsenic content of the lungs of the mice reared in a dust-free atmosphere was 0·36 parts per million. For the mice reared in a normal atmosphere, the corresponding figure was 0·82 parts per million.

Arsenic in detergents

Excitement is aroused from time to time by the discovery (or rather rediscovery) of arsenic in detergents. Many detergents include components made by the treatment of hydrocarbons with sulphuric acid. Some sulphuric acid is still made by the lead chamber process from pyrites, which often contains arsenic as an impurity. The presence of arsenic at relatively high levels (up to 74 ppm) in household and laboratory detergents was first reported in 1958.

Routine analysis of hair samples showed abnormally high arsenic levels, up to 42 ppm, in a number of female laboratory workers in Glasgow.

Washing of the hair samples to remove suspected external contamination did not alter the arsenic levels. It was then found that the detergents used to wash the hair samples contained substantial amounts of arsenic—and that the laboratory workers had been in the habit of borrowing the same detergents from the laboratory for domestic use. The high arsenic levels found during this investigation were eliminated by changing the source of supply of sulphuric acid used by the detergent manufacturer. Many commonly used detergents still contain arsenic at levels of a few parts per million or less.

The presence of arsenic at these concentrations is probably not harmful, but may be partly responsible for the higher arsenic levels found in women's hair than in men's hair, since men do not wash their hair as often as women. It is, of course, also significant in this connection that men spend more of their time in smoky surroundings.

Occupational poisoning

Arsenic is used, though to a decreasing extent, in the manufacture of insecticides (lead arsenate), weed-killer (sodium arsenite), and sheep-dip (arsenious oxide, sodium arsenite). Precautions and hygienic measures to avoid personal contamination are not always observed. A worker in a sheep-dip factory was admitted to hospital suffering from squamous carcinoma of the scrotum; the discovery (by Percival Pott) in 1775 of the prevalence of this tumour in chimney sweeps marked the beginning of the study of cancer as an occupational hazard. The patient just mentioned had certainly been exposed to arsenic in considerable amounts. Activation analysis showed the following arsenic levels, on admission to hospital:

Head hair	329 ppm
Finger nail	117 ppm
Skin	1·86 ppm

Samples of beard hair, obtained with an electric razor, showed the following arsenic levels:

On admission	3·12 ppm
7 days after admission	1·79 ppm
21 days after admission	0·84 ppm
28 days after admission	0·94 ppm

Beard hair is a useful material, since it can easily be sampled every day. The observation that excretion by this route falls to normal levels in about three weeks after cessation of exposure to arsenic, has been confirmed in several subsequent experiments.

Attempted suicide

The usefulness of the built-in time scale carried by every hair was illustrated in a case of arsenic poisoning studied in Glasgow in 1970. The subject swallowed several grams of arsenious oxide—many times the fatal dose—but vomited shortly afterwards and survived. A single hair, taken 49 days later, was cut into three-millimetre lengths, and the arsenic level in each section was measured by activation analysis.

The fifth section from the root showed an arsenic level of 94·4 parts per million; other sections showed much lower levels. Since hair grows at the rate of about 1 centimetre per month, this finding was consistent with the date of the reported ingestion of arsenic.

Mercury hazards in dental practice

The prevalence of mercury poisoning (usually mild though occasionally fatal) in dentists and their surgery assistants provides an interesting example of a hazard which, though widespread and clearly undesirable, is not yet adequately controlled. The hazard arises from the use of amalgam fillings—a practice which is more than 150 years old. The material commonly used to fill cavities in teeth is an amalgam of mercury with a silver-copper alloy. This material, apparently discovered empirically, has excellent thermal and mechanical properties and has never been seriously challenged by alternative preparations.

The use of amalgam fillings presents no hazard to the patient. Metallic mercury passes through the digestive tract with little or no absorption. Amalgam is almost insoluble in saliva and gastric juice. Increased excretion of mercury has been demonstrated (by activation analysis of urine) for a few days after the insertion of amalgam fillings, and measurable amounts of mercury (also detected by activation analysis) are found in the enamel layers of teeth adjacent to amalgam fillings—but there is no evidence that the small amounts of mercury absorbed in these ways are harmful to the patient. However, a patient may attend only once a year or less, whereas a dentist may insert twenty or thirty amalgam fillings in a day.

The dentist and his assistants are particularly vulnerable because of the techniques by which the amalgam is prepared for insertion. It was at one time common to amalgamate the constituents manually, using a pestle and mortar. More recently, use has been made of mechanical amalgamators, in which pre-filled capsules containing appropriate amounts of material are agitated vigorously to produce a pellet of the required size and composition. During this process the amalgam naturally becomes warm, and a quantity of mercury vapour is inevitably released when the capsule is opened.

To produce a filling of optimum mechanical properties, it is usually considered desirable to start with amalgam containing a slight excess of mercury, which is expressed by pressure between finger and thumb, sometimes with the intervention of a gauze or paper napkin. The environmental mercury level is further increased by evaporation from the filling, from spilled mercury and waste amalgam, from instruments sterilized by heat in the surgery, and from dust particles released when old fillings are drilled out. The common habit of smoking during intervals in the day's work also adds to the dentist's intake of mercury vapour.

Unmistakable signs of damage from mercury poisoning do not occur until the process is far advanced. The early symptoms, including minor degrees of irritability and tremor, are often plausibly attributed to the effects of age, overwork or alcoholic indulgence. It is, however, possible by activation analysis (and by no other method) to obtain a quantitative measure of mercury uptake before symptoms of damage become obvious. The analysis of head hair will, of course, reveal mercury deposited by internal and external pathways; mercury vapour from the atmosphere can pass through the outer layer of the hair and become firmly bound to the underlying keratin. Body hair will normally show only mercury which has been metabolized and deposited by the internal route. Similarly, finger nails will show internal and external contamination, but toe nails will normally show only internal contamination.

In a large study in Glasgow in 1973, samples of hair and nail were obtained from 87 dentists and 80 surgery assistants. Estimation of mercury by activation analysis showed considerably elevated levels (Table 4.2).

Table 4.2 Mercury levels in dental workers

	Mercury levels, ppm (geometric mean)			
	head hair	pubic hair	finger nail	toe nail
87 dentists	9.84	3.95	62.0	3.84
80 surgery assistants	9.53	3.02	17.3	7.52
16 other staff not handling mercury	3.52	1.59	3.51	1.58

Sensitivity of the analytical technique is illustrated by comparison of the mercury levels in the dentists (mostly male) and in the surgery assistants, all female. In head hair, pubic hair and finger nail, the mercury levels were higher in the dentists than in their assistants—but in toe nail, the surgery assistants showed considerably higher levels than the dentists. The explanation is that many of the surgery assistants wore sandals or other footwear with the toes exposed. Mercury vapour, sinking to the floor of the surgery, readily passes through the outer layer of the toe nail to become firmly bound and revealed by the subsequent analysis.

Surveys conducted in Scotland since 1973 have all given similar results, suggesting that between 5 and 10 per cent of all dentists and surgery assistants are absorbing excessive amounts of mercury vapour. The exact stage in the process of preparing and inserting the amalgam at which the dentist and his assistants are vulnerable is not yet known. There is no significant correlation between mercury uptake (as indicated by hair and nail levels) and work load, method of preparing amalgam, atmospheric mercury levels, or other factors which might be thought relevant. Meanwhile it is obviously desirable that the regular analysis of hair samples (at six-monthly intervals) should be instituted as a monitoring regime, so that staff who are at the greatest risk from mercury uptake may be identified before the onset of serious damage. A monitoring service would also assist in the development of safer working procedures.

Environmental lead

Lead is an element which requires—and receives—intensive study because the amount being put into the environment by human activity (notably from tetra-ethyl lead added to petrol to improve its properties as a motor fuel) has increased hugely during the past few decades. There is a belief, not yet supported by conclusive evidence, that children are at risk of brain damage from exposure to lead in the environment, and there is also anxiety over the health of workers occupationally exposed to lead dust or vapour. Individual uptake of lead is often monitored by the analysis of blood samples. Measurements on blood reflect short-term exposure and are therefore of limited usefulness, though probably reliable when exposure to lead is uniform and continuous.

The estimation of lead in hair (which is easily done by atomic absorption) gives a useful indication of integrated exposure over a period of time. Hair is more sensitive than blood as an indicator of changes in lead intake. The concentration of lead is usually about a hundred times higher in hair than in blood, and accurate measurement is correspondingly easier.

The usefulness of hair as an indicator of lead intake is illustrated in the results of a survey recently carried out in Glasgow and in East Kilbride, a nearby town with no heavy industry. Hair samples were taken from three groups of school children:

(a) those living in Glasgow in houses with lead water piping and/or lead storage tanks,
(b) those living in Glasgow in houses without lead plumbing,

(c) those living in East Kilbride, where no houses have any lead plumbing.

The lead levels in hair (Table 4.3) show unmistakable differences among the three groups. The study of lead levels in hair would be a useful method of assessing the effect on lead intake of industrial or motorway pollution.

Table 4.3 Lead in children's hair

Residence	Lead in hair, ppm (geometric mean)
Glasgow, lead plumbing (89 children)	53·9
Glasgow, copper plumbing (104 children)	28·9
East Kilbride, copper plumbing (81 children)	10·2

Because of sampling difficulties it is unlikely that large-scale surveys of this kind could ever be done by the estimation of blood lead; hair is, however, very easily sampled, stored, transported and analysed.

Analysis of hair is useful in studying the consequences of environmental disasters. During the winter of 1971/72, several thousand people in Iraq developed mercury poisoning after consuming grain which had been treated with fungicides containing methyl mercury. The grain was intended for use as seed, but considerable amounts were used to prepare food. Hair samples collected during the spring of 1972 (about 3 months after exposure) were examined in Baghdad by activation analysis. The mercury levels in hair were found to be well correlated with severity of symptoms. The normal levels in unexposed persons were under 3 ppm. Among those who ingested methyl mercury but showed no symptoms, mercury levels in hair had values up to 300 ppm. Mild symptoms of poisoning (slight tremor and blurring of vision) were associated with hair mercury levels between 120 and 600 ppm, moderate symptoms (partial paralysis, difficulty in hearing) were observed in people with hair mercury levels between 200 and 800 ppm, and those showing severe symptoms (paralysis, blindness, deafness, loss of speech) had hair mercury levels between 400 and 1600 ppm.

After recent volcanic activity in the Westerman Islands, near Iceland, atmospheric mercury levels were found to be increased. Hair samples obtained from 15 islanders were examined in Glasgow by activation analysis. Mercury, arsenic, selenium and zinc were all present at normal levels and were uniformly distributed along the hairs, indicating that no abnormal intake had resulted from the volcanic eruptions.

Environment and disease

There is a good deal of evidence that some of the major problems of medicine, including heart disease and cancer, have their origins in the environment. The World Health Organization has a substantial research programme involving the analysis of hair and nail samples from many parts of the world to test the hypothesis that variations in the concentration of trace elements (as yet unidentified) in the environment are linked with the differing incidence of hypertension and other cardiovascular diseases in different parts of the world. So far no significant results have emerged from these studies. Another WHO project involves the analysis of hair and nail samples to see whether environmental trace-element levels are associated with the striking geographical differences in cancer prevalence. This work is also still at an early stage.

The historical environment

Hair has the unique ability to retain the image of the environment, imprinted during life, for a very long time thereafter. The chemical composition of hair makes it resistant to most forms of decay. Consequently hair samples hundreds of years old can be found in excellent condition, apparently little changed by the passage of time. Hair which has been stored above ground is naturally more suitable for analysis, but material which has been buried in a sealed container survives well. Roman hair, nearly 2000 years old, found recently during an excavation at Dorchester, was analysed by activation analysis and found to contain mercury and arsenic, at about the levels common today. Thirteenth-century hair, from a burial site at Sweetheart Abbey in Dumfriesshire, contained 42 ppm of mercury—a remarkably high value, almost certainly indicating medicinal exposure.

Samples of hair from eminent individuals, preserved for historical or sentimental reasons, sometimes give useful information about an individual's chemical environment, internal or external. A number of cases of this kind have been studied in recent years in Glasgow.

Anne Mowbray, who was married as a child to Richard, Duke of York (one of the Princes in the Tower) died in 1481. Her remains were found in a lead casket on a building site in Stepney in 1964. Hair samples, examined in Glasgow by activation analysis, were found to contain 3·3 ppm of arsenic and 9·1 ppm of mercury. These values are considerably higher than would be expected today in the hair of unexposed persons, where normal values are 0·1–1 ppm of arsenic and 0·5–3 ppm of mercury.

One hair was divided into 15 one-centimetre lengths. The arsenic contents of successive pieces varied considerably, suggesting repeated intake of arsenic, no doubt for medicinal purposes.

Charles II, King of England from 1660—1685, was an enthusiastic amateur scientist and was the founder of the Royal Society. He had a laboratory at Whitehall, where he experimented on the distillation of mercury—a favourite pastime of the alchemists; according to Macaulay he showed more enthusiasm for his scientific recreations than for the affairs of state. The memoirs of the King's physicians suggest that he died of uremia, following kidney failure—a common feature of mercury poisoning. Two American physicians suggested in 1961 that he had indeed died of chronic mercury poisoning, as a result of inhaling excessive amounts of mercury vapour.

A sample of the King's hair, examined by activation analysis in 1966, showed 53 ppm of mercury. This is about twenty times the normal level in an unexposed person, and certainly supports the conclusion that the King had absorbed considerable amounts of mercury. Whether this was the cause of his death is another matter, which could be decided only by the analysis of further samples of known date, including some taken near the end of the King's life.

The medical history of Scotland's national poet has been studied inconclusively for many years, with rheumatic fever as the most favoured diagnosis. An unusual possibility was suggested in 1844 by John Thomson, a physician who, as a boy, had been a friend of the poet for a few years before his death in 1796. Thomson (in a little book entitled *Education: Man's Salvation from Crime, Disease and Starvation; with Appendix vindicating Robert Burns*) wrote as follows:

Fame, prompted by priests yet countenanced by friends, has promulgated an untruth that Robert Burns died, prematurely died, dissipation's martyr. From personal correct knowledge I proclaim that Robert Burns died the doctor's martyr...
The truth stands thus—the physician of Robert Burns believed that his liver was diseased, and placed him under a course of mercury. In those days a mercurial course was indeed a dreadful alternative. I know well that his mercurial course was extremely severe...
Among the last words I ever heard him speak were, "Well the doctor has made a finish of it now".

A sample of Burn's hair, examined by activation analysis in 1971, showed 8·0 ppm of mercury. This is well below the levels associated with mercury poisoning, but is higher than would be expected in a normal subject today. Since dental amalgam and agricultural chemicals did not contribute to the mercury content of the environment in Burn's time, the finding of 8 ppm confirms the supposition that Burns had absorbed mercury. This is not surprising, for mercury was widely used in medicine in the eighteenth century—and indeed until well into the twentieth century.

One of the most interesting patients studied in retrospect by activation analysis is Napoleon. It is generally believed that his death in 1821 (while a prisoner of the British on St. Helena) was due to cancer of the stomach. In 1962 Dr Sten Forshufvud, a Swedish dentist and historian, suggested that the Emperor had died of arsenic poisoning. A sample of Napoleon's hair, taken shortly after he died, was found on examination by activation analysis to contain 10·4 parts per million of arsenic—about 15 times the level to be expected in an unexposed person today.

Further samples of Napoleon's hair, all with reasonable proof of authenticity, were obtained from a variety of sources, and all showed substantial arsenic levels (Table 4.4). It is significant that the highest levels

Table 4.4 Arsenic content of Napoleon's hair

Date of Sample	As, ppm
1816	56·1
	38·7
	48·6
	19·4
1817	3·2
1818	21·2
	7·5
1821	10·4
	4·9
	3·8
	3·3

were found in samples given to Commander John Theed, R.N., when his ship called at St. Helena on 14th January 1816—three months after Napoleon arrived there. These samples must have reflected the Emperor's exposure to arsenic before he was captured.

A single hair, some 4 centimetres long, taken from Napoleon's head just after his death, was irradiated and then cut into 1-millimetre pieces before the arsenic determination was made. The results of this analysis (figure 4.1) are consistent with regular administration of arsenic, presumably for medicinal reasons. Arsenic, like mercury, was widely used in medicine during the nineteenth century.

Findings on hair samples from characters in history are, in themselves, mere anecdotes. Work of this kind is, however, important for the light that it throws on the internal chemical environment resulting from exposure to toxic metals in earlier times. The medicines available to the rich and famous were, in general, the same as those given to the rest of the population; until the beginning of the present century, there were rather few drugs available to the physician, and many of them (such as arsenic and mercury) did more harm than good.

Further evidence of the general state of the internal chemical environment has been obtained by the analysis of a large number of hair samples

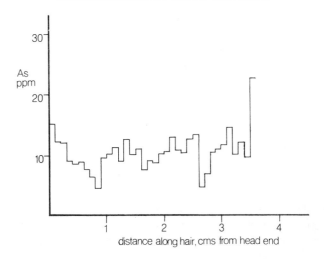

Figure 4.1 Arsenic content of one of Napoleon's hairs, taken on 6 May, 1821 (Forshufvud, S., Smith, H. and Wassen, A., (1964) *Archiv für Toxikologie*, **20**, 260).

Table 4.5 Mercury and arsenic in human head hair: historical and present-day levels

	ppm (geometric mean)	
	As	Hg
19th century (26 samples)	3·77	4·71
1973–74 (82 samples)	0·13	2·41

from people who lived during the nineteenth century and during the early part of the twentieth century. In many Victorian families, small locks of hair were removed from the dead, stitched to ornamental cards giving the name, date of death and age of the deceased, and distributed as momentos to friends and relatives. Many of these keepsakes have survived, and small fragments have been analysed in Glasgow. These experiments (Table 4.5) show that the general level of internal contamination by toxic metals has diminished considerably during the past century and that the internal environment is not (as is often supposed) becoming more polluted through thoughtless industrial or agricultural activity.

It is not always appreciated that great amounts of toxic substances were used for medicinal purposes in earlier times. References to any nineteenth-century textbook of materia medica or pharmacy shows that heroic amounts of arsenic, mercury, lead and antimony were prescribed for a great variety of illnesses. The early editions of Mrs Beeton's *Book of Household*

Management, used by almost every Victorian housewife in Britain, contains homely hints on the care of children. Mrs Beeton recommended compounds of mercury and antimony (as well as other toxic substances) for the relief of almost every childhood ailment; hair from a four-year-old girl who died in 1877 contained 286 ppm of mercury. The decline in the use of toxic metals and their compounds in medicine has been a major factor in the purification of the internal chemical environment.

There is, however, no justification for complacency. The natural cycles which have for thousands of years regulated the passage of toxic metals through the biosphere are, in many instances, being seriously disturbed by increased industrial activity and other human activities. Though arsenic, mercury and lead have received much publicity in recent years, many other trace metals, on which we are so far not well informed, may be linked with disease, even when present at levels not now considered to be dangerous. A great deal remains to be done to achieve satisfactory understanding of the role of trace elements in relation to nutrition and disease. In this work the analysis of hair offers many advantages and deserves to be more widely exploited.

FURTHER READING

Schroeder, H. A. (1976), *Trace Elements and Nutrition*, Faber.
Various papers in the annual series *Trace Substances in Environmental Health*, University of Missouri, Columbia, Missouri.
Hammer, D. I. (1971), "Hair Trace Elemental Levels and Environmental Exposure," *Amer. J. Epidemiol*, **93**, 84.

CHAPTER FIVE

DRUG TOXICITY

J. P. GRIFFIN

THE MAGNITUDE OF THE PROBLEM

The problem of adverse reactions to drugs is of broad scientific, political, economic and moral concern. This concern was reflected in a recent United States Congressional hearing where testimony was given that

it is now known that billions of wasted dollars, hundreds of thousands of unnecessary hospitalizations for adverse drug reactions, and thousands of lives needlessly lost are the price that society pays for the promotional excesses of the drug industry.

Melmon (1971) stated that the economic consequences of drug reactions are staggering; one seventh of all hospital days is devoted to the care of drug toxicity at an estimated yearly cost of $3 000 000 000.

The gathering of data

The gathering of valid data to support such statements is beset with problems. The reported incidence of reactions in various surveys is variable, reflecting the different methods of surveillance, the populations under study, and the habits of medical care. In addition, there are surprisingly few reports in the literature on the incidence of drug reactions in general, hospital or specialized practice. The studies that have been published in recent years are, however, enlightening and informative.

In the Johns Hopkins Hospital, between 1 January and 31 March 1964, 184 adverse drug reactions were seen in 122 patients out of a total of 714 admitted to general medical beds. Drug reactions following admission were seen in 13·6 percent of patients, and reactions present on admission were seen in 5 percent of patients and were the major cause for their admission. There were six fatal hospital-acquired adverse reactions (Seidl et al., 1966).

The same group of workers (Smith, Seidl and Cluff, 1966) also investigated the incidence of drug-induced adverse effects in a 33-bed medical ward between 1 January and 31 December 1965.

Nine hundred patients were included in the survey, and reactions were detected in 10·8 percent of patients. Reactions were most common in patients who were seriously ill and who had received many drugs. Patients with abnormal renal function or with previous drug reactions appeared to be predisposed to drug reactions, and allergic reactions were common in patients with gastro-intestinal disease. Altogether 133 drugs were incriminated in 114 adverse drug reactions: barbiturates, meprobromate, codeine, penicillins and thiazide diuretics produced 70 percent of the adverse reactions, although they accounted for only 17 percent of all medication.

In 1967 Ogilvie and Ruedy surveyed adverse drug reactions in 731 patients admitted to a general medical unit over a twelve-month period, and 193 patients (18 percent) suffered undesirable consequences of drug therapy.

One-quarter of the 67 deaths in the unit were due to adverse drug reactions. The majority of the reactions was caused by drugs that had been in common use for many years; indeed 60 percent were caused by digitalis, antimicrobials, insulin and diuretics. Most of the reactions (81 percent) were caused by the primary pharmacological action of the drug, or overdosage, or side-effects, or cytotoxic effects; these reactions were dose-related and predictable. Nineteen percent of reactions were due to interaction of the drug with special predisposing factors that were constitutionally induced, disease-induced, drug-induced or environmentally induced.

Data available from the Boston Collaborative Drug Surveillance Program provide a resource that can be used to determine with considerable accuracy how often drugs are used and how toxic they are. The data base of the Boston Collaborative Drug Surveillance Program, which has been in operation since 1966, allows for the evaluation of drug effects during hospitalization into medical departments. The short-term effects have been studied in 19 000 in-patients admitted to medical wards. The information about out-of-hospital reactions derives from the study of 40 000 patients.

Jick (1974) assessed the magnitude of the problem of adverse reactions to drugs in the United States as follows.

About 30 percent of hospitalized medical patients have at least one adverse reaction during hospitalization and, using the estimate that one third of the approximately 30 million people admitted to United States general hospitals are admitted to general medical services, we estimate that 3 000 000 hospital patients suffer an adverse drug reaction in medical units each year. Similarly, taking our estimate that about 3 percent of admissions to these services are due to adverse drug reactions, and applying this to the total annual number of patients admitted, it is apparent that hundreds of thousands of people are admitted each year because of adverse drug reactions. Finally, our estimate (derived from some 11 500 monitored patients) that drug-attributed deaths occur in 0·29 percent of hospitalized medical patients implies that the total annual number of deaths in this country (the USA) is about 29 000.

Each patient in the Boston Drug Surveillance Program monitoring

DRUG TOXICITY 89

scheme received an average of nine drugs whilst in hospital and, allowing that 30 percent of patients treated in hospital showed an adverse reaction, Jick estimated that in only 5 percent of the courses of drug therapy are there adverse reactions.

In the case of patients treated outside hospital, Jick estimated that in the United States 75 million adults took an average of 2 drugs regularly and that 300 000 hospital admissions per year resulted from this usage. Overall hospitalization occurred for one out of every 500 long-term drug treatments per year. The highest rates were seen in patients receiving digitalis, glycosides, anti-tumour drugs, steroids and anticoagulants.

The overall conclusion that was drawn in this review was that, although the problem of adverse reactions to drugs was numerically large, it had to be appreciated that, in view of the widespread and extensive use of drugs therapeutically, the hazard from a drug to individual patients was an acceptable risk.

It is difficult to reconcile the relatively high incidence of drug-induced reactions obtained in the quoted individual studies with official figures published on the incidence of adverse drug reactions.

The Australian Drug Evaluation Committee in their Report on Adverse Drug Reactions recorded a total of only 628 reactions during the period October 1965 to December 1967, which was only 3 per 100 000 of the population at risk. Furthermore, during 1968 the Committee on Safety of Drugs received a total 3446 notifications of adverse reactions occurring in the United Kingdom, during which period approximately 306 000 000 prescriptions were issued.

Medicine in the Public Interest Inc., a non-profit corporation studying issues related to medicine and science, released a lengthy drug reaction report in December 1974. In this report the following views were propounded.

Estimates of the magnitude and cost of the adverse drug reaction problem are unreliable because of incomplete, unrepresentative, uncontrolled and statistically deficient data. Other objections raised were that minor reactions such as gastro-intestinal disturbances, rash, itching, drowsiness, insomnia, weakness, headache, muscle twitches and fever account for 71 percent of reported adverse reactions; and many adverse reactions occur in gravely ill patients with underlying disease.

These comments cannot, of course, be used to object to the estimates of the incidence of drug toxicity made by Jick, and in fact only reiterate an assessment made two years earlier (D'Arcy and Griffin, 1972).

Two factors are the major contributors to the manifestation of iatrogenic disease. These are the abnormal patient reaction to the drug, and the development of unexpected toxicity when several drugs are given in combination.

The most urgent need in the field of control of drug toxicity in the interests of patient safety is to determine not only the adverse reactions that a drug may cause, but the incidence of those reactions in relation to the use of the drug, and to determine sections of the population at greater than average risk.

Some of the drugs involved

The drugs that cause major problems with respect to adverse reactions are among the best known and most widely used, e.g. digitalis, penicillin, insulin, hypnotics.

In a survey in Belfast it was found that 19·8 percent of patients treated with digoxin suffered adverse reactions to it, and that adverse reactions to digitalis accounted for one third of all drug reactions monitored.

Withering in 1785 was aware of the fact that the drug he had pioneered was not without hazard and wrote:

it is better the world should derive some instruction, however imperfect, from my experience than that the lives of men should be hazarded by its unguarded exhibition, or that medicine of so much efficacy should be condemned and rejected as dangerous and unmanageable.

Another group of drugs reported to give rise to a high incidence of side effects was the oral anticoagulant drugs. This was investigated by Williams, Griffin and Parkins (1975) in a series of 277 patients, each of whom was on long-term anticoagulants, and each patient was seen monthly for a period of six months.

In this series, during the six-month period when 413 courses of treatment with other drugs were given to these patients, a theoretical drug interaction with the anticoagulants could have occurred with 94 of these courses of treatment. In actual practice 33 adverse reactions were recorded, of which 6 were due to drug interactions and one to superimposed lead poisoning interacting with the anticoagulant. The overall adverse reaction rate to the anticoagulants was therefore 12·1 percent.

Even with simple drugs such as potassium chloride, the adverse reactions are not inconsiderable.

Lawson (1974) studied 16 048 patients admitted consecutively to the Boston Drug Surveillance Program; 4921 (31 percent) received potassium chloride. The major indication for the use of this drug was prophylaxis against electrolyte depletion in patients receiving diuretic therapy. Adverse reactions attributed to potassium chloride were reported in 283 (5·8 percent) patients, the most common reaction being hyperkalaemia. Adverse reactions culminated in the death of 7 patients and was life-threatening in a further 21 cases.

In respect of the use of potassium chloride supplements, data are now accumulating that in many circumstances in which the preparation is prescribed it is being prescribed unnecessarily and inappropriately.

THE NATURE OF THE PROBLEM

As a result of epidemiological studies on adverse drug reactions, it can be stated that in a breakdown of the problem the following factors must be considered.

(a) Pharmacological overdose.
(b) Inherent drug toxicity, unrelated to pharmacological actions.
(c) Interaction between two or more drugs to produce an untoward and unwanted reaction.
(d) Patient factors, such as genetic differences in drug metabolism or disease.

In the following pages these aspects will be discussed using individual problems as examples. However, pharmacological overdose will be omitted, as this is self-explanatory.

Inherent drug toxicity

Adverse reactions to drugs themselves may occur due to pharmacological overdose or to toxic reactions unrelated to the pharmacological actions or desired therapeutic effect of the preparation; and these effects should be considered separately from adverse reactions due to atypical patient factors, or to drug interaction. The adverse reactions so produced may not be due to the active ingredient itself, but to other excipients included in the preparation. All too often it is forgotten that it is not a simple drug, but a formulation containing other ingredients, such as antioxidants, preservatives, solubilizing agents and colouring materials, that is administered to the patient. Instances can be cited where drug toxicity was due to the excipients rather than to the active principle; e.g. cataract formation seen with preparations containing dimethyl sulphoxide (DMSO) as a solvent.

Apart from toxicity due to pharmacological overdosage, amongst which the commonest are overdosage with digitalis, glycosides and anticoagulant drugs, toxicity unrelated to the pharmacological action of the drug is comparatively rare, but for obvious reasons receives considerable attention.

It was in 1877 at a meeting in Manchester that the British Medical Association initiated the first collaborative study of adverse reactions to a drug. A committee was set up to investigate the sudden and unexpected deaths which sometimes occurred during the induction of chloroform anaesthesia. At the end of the First World War, an epidemic of jaundice and fatal hepatic necrosis amongst soldiers treated for syphilis with organic arsenicals was so serious that it was the subject of a special report by the Medical Research Committee, predecessor of the present Medical Research Council.

The association between drug exposure and drug toxic reaction is easier if the time interval between the two is short; and the longer the time interval between the two events, the more difficult it is to establish cause and effect.

Instances where the interval between drug exposure and the manifestation of an adverse reaction is short, and therefore cause and effect are easy to establish, are drug-induced rashes, overdosage with digoxin, or oral anticoagulants, or drug-induced parkinsonian tremor with chlorpromazine-like antidepressants. Examples where the latent period between exposure and toxicity is long can be seen with respect to the cancer-producing action of some drugs, or the adverse effects of drugs on the developing foetus. The first suggestion that the therapeutic administration of arsenic could lead to cancer was made by Hutchinson (1887) who described five cases of skin cancer following the medicinal use of arsenic. It is now generally accepted that arsenic given by mouth can cause cancer of the skin, which is usually preceded by arsenical pigmentation, keratosis and dermatosis. Fierz (1965) found that, in a series of 262 patients treated with Fowler's solution for from 6 to 26 years, carcinoma of the skin developed in 21 patients (8 percent). The type of cancer most frequently encountered was a basal-cell carcinoma, and in 16 of these patients there were multiple basal-cell carcinomata.

Sommers and McManus (1953) described a series of 27 cases of arsenical skin cancer, and drew attention to the fact that 10 of these patients also had other primary sites of neoplastic change. Two cases had bronchial carcinoma, and three had primary carcinomata arising in the genito-urinary tract. This association of arsenic with visceral carcinogenesis was further elaborated by Robson and Jelliffe (1963) who described six cases (four female and two male) all of whom developed bronchial carcinoma with an average latent period of 32 years after receiving arsenic therapy. Arsenic had been prescribed to these patients for psoriasis, rheumatic fever, "convulsions", or as a "tonic". Each of these patients had the dermatological stigmata of arsenic ingestion.

The major importance of discussing arsenic-induced cancers now is that owing to the long latent period of onset of clinical neoplasia such patients may still present themselves for treatment. It should also be borne in mind that, although there are no indications today for using arsenic, the organo-arsenicals have been used in chemotherapy, and tryparsamide, which contains about 25 percent of pentavalent arsenic in organic combination, is still used in the treatment of African trypanosomiasis.

Several aspects of inherent drug toxicity will be reviewed individually.

The thalidomide problem

It was the publicity surrounding the thalidomide disaster that brought the problems of inherent drug toxicity to public attention and had resounding effects in terms of legislation. In view of its historic importance it might be desirable to consider the spectrum of thalidomide toxicity in some detail.

The first adverse reaction reported to be caused by thalidomide was the least important—the development of myxoedema. This report was made in 1959.

Early in 1960 isolated reports were received by Burley of Distillers Company (Biochemicals) Limited, from various parts of Great Britain, describing symptoms and signs suggestive of peripheral neuritis occurring in patients receiving thalidomide regularly for periods of six months or more. Florence, however, recorded the first report in the literature in December 1960. In four patients polyneuritis had developed while they were taking thalidomide, and he thought that the symptoms could possibly be a toxic effect of the drug. Kuenssberg *et al.* soon added five similar cases in January 1961. It was not till the more detailed report of Fullerton and Kremer in September 1961 that the association of thalidomide and resulting neuropathy became fully realized in Great Britain.

In December 1959 Weidenbach presented the case history of a girl born on November 10, 1958 to a twenty-four-year-old primigravida. The upper and lower limbs were missing. The hands and feet originated directly from the shoulder and pelvic girdle respectively. There were also deformities of the digits. No additional abnormalities were noted. The infant progressed very well in the nursery and continued to develop in accordance with her age. The history of the gestation yielded nothing unusual. Neither parent could recall a family history of malformation. Although it was recognized that no conclusion regarding the aetiology of the malformation could be drawn, it was thought that, owing to the symmetry and involvement of all extremities, a hereditary factor was most likely.

Kosenow and Pfeiffer, at the September 1960 meeting of the German Society of Paediatrics in Kassel, had a scientific exhibit describing two infants with similar malformations and also micromelia, hemagioma of the midline of the face and duodenal stenosis.

In September 1961, Wiedemann presented a paper calling attention to the current increase in the incidence of hypoplastic and aplastic malformations of the extremities. Over a period of ten months he had seen 13 patients. He was aware of 27 similar cases in his area. Since no hereditary signs appeared in the histories of any of his 13 patients, he considered an exogenous cause that must have come into effect around the beginning of 1959. He questioned whether a drug among the constant flow of new drugs entering the market might have been involved.

Pfeiffer and Kosenow presented a paper on the question of exogenous causes of severe malformations of the extremities to the North Rhein-Westphalia Paediatric Meeting in Dusseldorf on November 18, 1961. They mentioned 34 newborn infants with defects of the long bones seen at the Children's Hospital at Muenster from January 1, 1960 to November 18, 1961.

The Fetal Life Study received an inquiry in 1962 concerning the incidents of phocomelia in relation to a recent increase that had been observed in West Germany. The Fetal Life Study was established in 1946 in a selected population at the Columbia-Presbyterian Medical Center as a long-term prospective epidemiologic investigation of human reproduction, to determine the incidence of foetal deaths, neonatal deaths and congenital malformations, and to delineate associated factors. From more than 10 000 pregnancies prospectively followed in the years 1946 to 1960, and more than 2000 followed in 1961, the group was unable to find any causes of phocomelia similar to the pictures appearing in the literature. A possible explanation of this discrepancy became apparent at the Rhein-Westphalia Paediatric Meeting in Dusseldorf on November 18, 1961. Lenz of Hamburg suggested that this malformation was related to the ingestion early in pregnancy of the drug thalidomide (alpha-(N-phthalmido)glutarimide).

It was not long, however, before individual case reports began to appear. These cases also illustrated the problems of retrospective epidemiology. In one situation the drug had to be retrieved from a former neighbour. In another case, in which the mother had been included in a hospital-study prospective survey, the fact that she was given thalidomide was not known by her family doctor. In retrospective studies, two out of three family doctors could not remember whether thalidomide had been taken.

Several communications indicated that small doses might be devastating. One patient who was a week overdue for a menstrual period took 50 mg of thalidomide a day for one week only; her premature baby had phocomelia. In another report the mother apparently received 100 mg of thalidomide for three nights and 50 mg for two nights in the second week of pregnancy; the baby was born with phocomelia. If the dates in these situations were correct, these may illustrate the earliest stages of pregnancy in which teratogenic effects should be sought in the assessment of teratogenicity. It is even more worrying that in these cases the drug was exerting its teratogenic effect in women who did not even know for certain that they were pregnant.

The continual appearance of new hazards and new dimensions of adverse reactions—diethylstilboestrol and adenocarcinoma of the vagina

The first confined reports of the transplacental transmission of cancer in man by means of a hormone, diethylstilboestrol, have recently been published. The evidence provided by this extremely important research and its significance need immediate and careful assessment.

In 1970 Herbst and Scully reported seven cases of adenocarcinoma of the vagina in adolescent girls in the New England area during a period of four years. The patients' ages ranged from 15 to 22 years. They had symptoms of irregular vaginal bleeding for up to one year. Five were treated by radical surgery and one by wide excision. All were alive one to four years after operation. The seventh, in whom the disease was too far advanced at surgical exploration, died within six months. The authors were puzzled about the causation of this apparent clustering of cases, as carcinoma of the vagina is uncommon and usually occurs at a much greater age.

An eighth case was added in a retrospective study of factors that might have been associated with the appearance of these tumours. Herbst and colleagues (1971) noticed that maternal bleeding when the girl's mother was pregnant with the patient (and in previous pregnancies) was more common than in a control group. But of greater significance than that was the finding that seven of the eight mothers had been treated with diethylstilboestrol during the first trimester of the material pregnancy, while none of the control group was so treated. A separate study has now confirmed this association, adding five more cases in which the actual dosage of synthetic oestrogen used has been obtained.

All 13 patients were born between 1946 and 1953, a period when diethylstilboestrol was being given for repeated or threatened abortion. All the mothers who took diethylstilboestrol began treatment in the first two months of pregnancy. They received either a constant dose administered throughout pregnancy or a continually increasing dose given almost to term. The actual dosage varied, but followed roughly that suggested by A. W. Smith, beginning at 5 mg by mouth during the sixth or seventh week of pregnancy, and increasing by 5 mg at two-weekly intervals to the 15th week, when 25 mg daily was being given. The dose then increased by 5 mg at weekly intervals until the 35th week, at which time as much as 125 mg of diethylstilboestrol was being taken by mouth daily.

The original series of seven cases exceeded the number of cases in the entire world literature for a tumour of this type in adolescent girls born before 1945. Indeed, adenocarcinoma of the vagina was thought to have some relationship to vaginal remnants. Moreover, if these neoplasms were the result simply of high-risk pregnancies, this should have become apparent before 1945. It was therefore suspected that exposure to diethylstilboestrol and vaginal carcinoma in the offspring might have a cause-and-effect relationship. The suggestion is reinforced by the fact that diethylstilboestrol was used only infrequently in general obstetric practice. Even at the Boston Hospital for Women, where a special high-risk pregnancy clinic was being conducted, only about 1 in 21 patients delivered in the wards had received diethylstilboestrol during the five-year period 1946–1951. Thus when the expectancy of a chance association is less than 5 percent, the occurrence of maternal diethylstilboestrol therapy in 12 out of 13 cases of vaginal carcinoma in young women cannot be considered coincidental.

In a Senate debate reported in the *Washington Post* (10th September 1975) Senator Kennedy stated that 220 cases of vaginal carcinoma were now known to have developed in the daughters of women treated during pregnancy with diethylstilboestrol.

Phenacetin renal damage—a problem of toxicity and drug abuse
The problem of renal damage with phenacetin containing analgesic preparations is another example where drug toxicity only becomes apparent after years of clinical use; it has been reviewed in detail by D'Arcy and Griffin (1972). Phenacetin was introduced into clinical medicine in 1887. The first report indicating that phenacetin when taken in high doses for considerable periods of time could damage the kidney was published in 1953. Over the past 20 years there have been numerous papers describing renal damage in the form of interstitial nephritis and necrosis of the renal papilla, and ultimately in some cases death from renal failure, in patients taking excessive amounts of phenacetin-containing preparations. The problem of phenacetin renal toxicity was given a new dimension by reports from Sweden in 1968 of cancer of the renal pelvis in patients with other renal damage known to have been caused by phenacetin.

Controls on the supply of phenacetin were introduced in the United States in 1964 and by most European countries at various dates over the following ten years, i.e. 77–86 years after the drug was first introduced as an analgesic drug.

Toxicity due to drug interaction

Today there is much concern about "drug interaction" because many patients receive more than one drug at a time. Many doctors are unaware of the risks to which their patients are exposed when treated with multiple drugs. When Osler, about 100 years ago, referred to the physician who practises

a sort of popgun pharmacy, hitting now the malady and again the patients, he himself not knowing which,

he little thought that his words would be applicable today. It has been pointed out that every time a physician adds to the number of drugs a patient is taking he may devise a novel combination that has a special risk hitherto unsuspected. Occasionally these risks are predictable on the basis of known pharmacology, but all too often they have emerged only after the exposure of many patients.

A drug interaction occurs whenever the presence of one chemical substance changes the pharmacological effects of a therapeutically administered drug. The term *chemical substance* in this context should be extended to include alcohol, foods, insecticides, possibly food additives, environmental chemical agents, as well as drugs therapeutically administered and drugs of abuse, such as cannabis and tobacco.

As a result of drug interaction, the pharmacological action of one or both agents may be modified, either by increasing or reducing the pharmacological action and the inherent toxicity of drug, or on the other hand may produce a totally novel toxicity when taken in combination.

Drug interactions can occur at the following sites (figure 5.1):

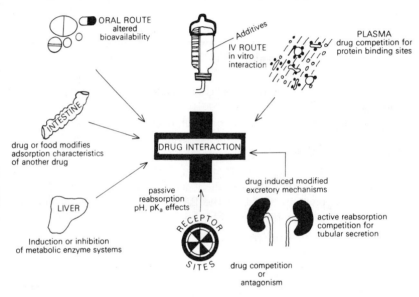

Figure 5.1 The possible sites of drug interactions.

(i) before administration. (These interactions are important when drugs are added to intravenous infusion fluids or mixed in syringes.)
(ii) in the gastro-intestinal tract (where a drug or food component may modify the absorption of another drug).
(iii) at protein sites in the plasma.
(iv) at sites of drug metabolism, notably the liver.
(v) at sites of drug excretion, notably the kidney.
(vi) at receptor sites in the tissues.

Drugs may interact at more than one site or by more than one mechanism, and it is often difficult to predict the end result. It must be stressed that the risk of an interaction does not necessarily prohibit giving the drugs involved in combination. Moreover, interactions may occur in some patients but not in others. With some drugs (e.g. monoamine oxidase inhibitors) serious interactions only seem to occur in a very small proportion of the patients at risk.

Drug absorption interference may arise in several ways. A change in the pH of the contents of the intestinal tract may alter the solubility of a drug, and thus decrease or increase the rate of absorption, e.g. sodium bicarbonate reduces the absorption of tetracycline. Some drugs interact directly in the gastro-intestinal contents to form complexes which may not be absorbed. The inhibition of tetracycline absorption by iron and antacid preparations containing calcium, magnesium and aluminium salts, or the impairment of warfarin absorption by cholestyramine, may be cited as examples of this form of interference with absorption.

One drug may influence gastro-intestinal motility and gastric emptying, and alter the rate of absorption of other drugs given at the same time. In general, drugs are poorly absorbed from the stomach, and absorption may be slowed if gastric emptying is delayed by anticholinergic drugs such as propantheline ("Pro-Banthine").

Many drug absorption interactions cannot be explained. Drugs for the treatment of tuberculosis are usually given in combination. However, sodium aminosalicylate (PAS) impairs the gastrointestinal absorption of isoniazid and rifampicin (with rifampicin, serum levels over an 8-hour period may be reduced by approximately a half; this effect on absorption is due to the excipient bentonite in the PAS). Nevertheless, isoniazid and rifampicin can be given together without interference with absorption.

Plasma protein binding. After absorption many drugs are extensively bound to plasma protein, particularly to albumin. Drug effects are related only to the free fraction. Important changes in the drug distribution can arise from competition between drugs for these protein binding sites. Certain groups of drugs seem to share a limited number of common binding sites, and one drug may displace another with adverse effects that are sometimes dramatic.

Drugs which can displace one another from plasma binding sites include oral anti-coagulants, non-corticosteroid anti-inflammatory agents (including aspirin), sulphonylureas, anticonvulsants and sulphonamides. Potentiation of the anti-coagulant action of warfarin by phenylbutazone and oxyphenbutazone is an important example of such an interaction.

Drug metabolism. The action of many drugs is terminated by their conversion to inactive metabolites by liver enzymes. The activity of these enzymes can be increased by prolonged administration of certain drugs (induction), the usual result being a reduction in the duration and magnitude of the drug effects. Barbiturates (especially phenobarbitone) can cause induction of drug-metabolizing enzymes, and may thus reduce the anticoagulant action of simultaneously administered warfarin by increasing the rate of its metabolism. After restabilization, stopping barbiturates will result in the warfarin re-achieving its full effect, which may precipitate a serious haemorrhage. On the other hand, one drug may inhibit the metabolism of another, and thereby increase and prolong its effects, e.g. azathioprine and mercaptopurine are converted to inactive metabolites by xanthine oxidase, and this enzyme is inhibited by allopurinol. It follows that the dose of azathioprine or of mercaptopurine should be reduced by about 75 percent if allopurinol is being given at the same time.

Renal excretion. One drug may interfere with the renal elimination of another, but for most drugs the kidney is not the main route of excretion, and few clinically important interactions of this type have been reported. Many drugs are extensively reabsorbed back into the circulation, as the tubular fluid is progressively concentrated in the nephron. The extent of this reabsorption may depend on the urine pH, and this is altered by sodium bicarbonate and carbonic anhydrase inhibitors such as acetazolamide ("Acetazide", "Diamox") or dichlorphenamide ("Daranide"). Acid urine favours the reabsorption of acidic drugs, such as phenobarbitone and salicylate, so that their renal clearance is reduced. The opposite effect is observed with basic drugs; the urinary excretion of quinidine and amphetamine is increased in acid urine and decreased in alkaline urine. (This principle forms the basis of the use of forced alkaline diuresis in the treatment of overdosage with phenobarbitone or aspirin.)

Many acidic drugs are secreted by a proximal renal tubular active transport mechanism, so that interactions may arise from competition for this system. Drugs actively secreted in this way include sulphonamides, penicillins, non-corticosteroid anti-inflammatory drugs, uricosuric agents, and thiazide diuretics. The well-known use of probenicid to slow down the urinary excretion of penicillin is an example of competition for tubular secretion, and thus of a drug interaction of this sort which happens to be clinically useful.

Receptor sites. Drugs may interact by antagonizing each other at the same receptor site (competitive antagonism) or at separate but physiologically related sites (physiological antagonism). Among many examples of competitive antagonism is that between histamine and antihistamines, and the reversal of morphine narcosis by nalorphine ("Lethidrone") and naloxan ("Narcan").

The adrenergic blocking drugs (guanethidine, debrisoquine and bethanidine) provide examples of a different type of drug antagonism. These antihypertensive drugs are concentrated at their site of action in adrenergic nerve endings by the same mechanism as that which takes up catecholamines. Chlorpromazine and tricyclic antidepressants, such as imipramine, desipramine, amitriptyline and nortriptyline, are inhibitors of this uptake process and may completely abolish the blood-pressure lowering effects of these drugs. Paradoxically, chlorpromazine may augment the blood-pressure lowering effects of other antihypertensive drugs.

The problem of adverse reactions to drugs caused as a result of drug interaction has been dealt with extensively by Griffin and D'Arcy (1975).

The patient factor in drug toxicity

Predictable toxicity is the manifestation of secondary pharmacological actions, and this is a hazard that can be well elucidated in the battery of tests to which the drug is subjected during its development stage. In such instances, assuming that the drug has a worth-while place in therapy, the ratio of dosage of drug to produce the major effect, to dosage to produce a secondary (toxic) effect is the real factor which should be considered, and not the "built-in" potentiality to the side-effect itself. Nevertheless, certain groups of patients exist who may be at risk from these predictable manifestations of toxicity. These patients are at peculiar risk because of a genetically determined defect of metabolism, or because their metabolism has been impaired by concomitant hepatic disease, or because their excretory function has been reduced by either liver or renal malfunction. In these patients, the drug or its metabolites may rapidly build up to toxic levels in the body, even at normal accepted therapeutic doses.

Intolerance is a lowered threshold to the normal pharmacological action of drugs. Individuals may vary widely from the well-established norm in their reaction to drugs. The very old and the very young are liable to be more sensitive to drugs, possibly because the metabolic and excretory mechanisms essential for the disposal of the drug are less efficient. In addition, the reactions of the old or the young may also differ qualitatively from those of the adult.

Adverse reactions may follow the use of a drug, and these reactions may be unexpected in that they are completely unrelated to the known toxicity of the drug. These reactions include hypersensitivity to the agent, in which the patient develops antibodies to the drug. The antigenic factor is usually a combination of drug with body protein. Skin rashes and eruptions are the most common symptoms of this type of allergic reaction, although haemolytic anaemia is not infrequent.

The patient factor in drug toxicity will be dealt with under the headings of genetic factors, age factors, disease factors and allergic factors. The predominant problem in this area of the field of drug toxicity is the identification of patients at special risk, and their protection from the possible adverse effects of therapy. This area has been neglected because of the greater stress which has been, perhaps mistakenly, given to the evaluation of drugs for their inherent toxicity.

Genetic factors

In some cases the genetic factor is such that the metabolism of the drug is delayed due to enzyme deficiency, and the primary pharmacological or secondary toxic effects of the drug are enhanced, e.g. plasma pseudocholinesterase deficiency or presence of atypical cholinesterases, and slow acetylators of isoniazid. On the other hand, the response of the patient because of these genetic abnormalities may be idiosyncratic. Idiosyncrasy involves a quantitatively abnormal response, an example of this being porphyria induced by barbiturates in susceptible subjects. Similarly, mepacrine-induced haemolytic anaemia in glucose-6-phosphate dehydrogenase-deficient subjects is an idiosyncratic reaction. In these patients with glucose-6-phosphate dehydrogenase deficiency, haemolytic anaemia may be induced by normal dietary constituents such as fava bean (*Vicia fava*) or even by inhalation of naphthalene vapour.

Fast and slow acetylation of drugs

Isoniazid, a drug used in the treatment of tuberculosis, is usually taken by mouth, and is completely and rapidly absorbed. Some patients achieved higher blood levels of isoniazid than others, and it was shown that the differences in blood levels were due to differences in the speed of acetylation and inactivation of isoniazid. Biopsy specimens from the livers of "rapid inactivators" of isoniazid had higher levels of hepatic N-acetyl transferase than biopsy specimens obtained from "slow inactivators".

The level of the N-acetyl transferase is genetically determined, and patients may be classified as "slow inactivators" (autosomal homozygous recessive) or "rapid inactivators" (heterozygous and homozygous dominants).

The relevance of the speed of acetylation of isoniazid in respect to therapy is twofold: (1) to achieve adequate therapeutic levels of the drug to treat the tuberculosis in the "rapid inactivator" and (2) to identify the slow inactivator who is more prone to the toxicity of isoniazid due to reduced ability to inactivate the drug. Slow inactivators are more prone to develop peripheral neuritis due to isoniazid toxicity and (to quote an example in a report of isoniazid-induced peripheral neuritis from the Tuberculosis Chemotherapy Centre in Madras) of 43 cases 36 (83·7 percent) were "slow inactivators". Central nervous toxicity manifested as grand mal epileptic fits were seen in six of these 43 patients, and other central nervous effects resembled encephalopathy or psychotic behaviour. All eight patients with central nervous effects of isoniazid were "slow inactivators".

Other drugs are metabolized by the same N-acetylation system, such as sulphadimidine; a sulphonamide antibacterial phenelzine; a monoamine oxidase inhibitor used as an antidepressant; the anti-leprosy drug, dapsone; and hydralazine. Phenelzine was shown to be a much more effective antidepressant in "slow inactivators" than in "rapid inactivators".

A need therefore exists to determine the "acetylator status" of patients before treatment with certain drugs, and a test has been devised. Each patient received sulphadimidine by mouth in a dose of 44 mg/kg body weight, and free and total sulphadimidine were estimated in blood, and urine collected six hours after treatment.

The findings of these investigators suggest that patients can be classified as slow inactivators of isoniazid if the proportion of acetylated sulphadimidine (total minus free) is (a) less than 25 percent in blood or (b) less than 70 percent in urine. They reported that the sulphadimidine test was easy to perform and that it had the additional advantage that the result could be determined on the same day as the test. Stored samples of urine, kept at room temperature for over a week, gave satisfactory results.

Using this type of test, epidemiological studies have been conducted. About 5 percent of Eskimos, 15 percent of Chinese, 60 percent of Asian Indians, 45 percent of American Negroes, 45 percent of Europeans, and 55–75 percent of Jews are slow inactivators of isoniazid, and isoniazid-induced peripheral neuropathy shows an incidence among these different populations in accordance with their acetylator status.

Another aspect of the problems caused in therapy by different acetylator status in different patients can be illustrated by the fact that isoniazid inhibits the metabolism of the antiepileptic drug phenytoin; and this slowing of phenytoin metabolism is more marked in "slow inactivators" of isoniazid, and can lead to phenytoin intoxication.

Familial abnormal response to neuromuscular blocking drugs

Suxamethonium (succinylcholine, diacetylcholine) was introduced in 1949 by Bovet and his colleagues. It is a potent neuromuscular blocking agent of the depolarizing type with very brief duration. This is because suxamethonium is rapidly hydrolyzed and inactivated by pseudocholinesterase present in the serum. It is used when only a brief period of muscle relaxation is needed, or it may be given by intravenous drip to produce prolonged neuromuscular blockade during surgery.

Within a few years of the drug's introduction, attention was drawn to the prolonged apnoea paralysis of respiratory muscles after using it in the presence of low plasma serum-pseudocholinesterase. A familial incidence of low pseudocholinesterase in the absence of overt disease was also described following suxamethonium apnoea in two brothers who had low plasma-pseudocholinesterase levels and both had, one of them repeatedly, a prolonged muscular paralysis following the injection of suxamethonium. Plasma pseudocholinesterase is responsible for the rapid destruction of suxamethonium, so that any reduction in pseudocholinesterase activity or plasma level of the enzyme prolongs the duration of the neuromuscular blockage. Production of pseudocholinesterase is depressed in certain pathological conditions, including liver disease, severe malnutrition, and hyperproteinaemia other than that due to renal disease. The action of the enzyme is inhibited by organophosphorus compounds, such as are used in nerve gases and insecticides. A commoner cause of a transient inactivation of pseudocholinesterase is the excessive administration by the anaesthetist (or surgeon) of drugs with anticholinesterase activity, such as cocaine, procaine, and lignocaine. Some ganglionic blocking agents, such as trimetaphancamsylate and phenactropinium chloride (phenacylhomatropinium chloride), are also powerful cholinesterase inhibitors.

There is, too, a genetically determined deficiency of the enzyme; and congenital complete absence of pseudocholinesterase has been reported.

Some people who are abnormally sensitive to suxamethonium have been shown to have an atypical form of pseudocholinesterase in their serum. This atypical enzyme hydrolyzes suxamethonium at a much slower rate than does the normal or usual type of serum-cholinesterase; consequently the apnoea which the drug induces is excessively prolonged.

Patients with an atypical form of pseudocholinesterase do not present any other recognizable abnormality, and they usually have prolonged apnoea following a single injection of suxamethonium. A technique has, however, been developed to detect the presence of this atypical enzyme by determining what is known as the "dibucaine number". This is the percentage inhibition of enzyme activity produced by the inhibitor dibucaine under certain standardized conditions. In the test, the activity of the enzyme (usual pseudocholinesterase, or atypical pseudocholinesterase) in the presence and in the absence of dibucaine (10^{-5}M) is measured by following the rate of hydrolysis of the substance benzoylcholine (5×10^{-5} M) spectrophotometrically at 240 nm. The reaction is carried out at pH 7·4 in phosphate buffer.

With this technique most people show a dibucaine number of about 80 (normal homozygotes); those with atypical pseudocholinesterase may be classified as dibucaine number 40–75 (heterozygotes), or dibucaine number 30 or below (abnormal homozygotes). The intermediate group (heterozygotes) have a mean dibucaine number of about 62 and are believed to synthesize both the "usual" and "atypical" forms of the enzyme. They do not generally show an abnormal response to suxamethonium.

In summary, current opinion is that prolonged apnoea after suxamethonium is due to quantitative defects of pseudocholinesterase activity which are genetically determined. Four allelic genes seem to control the inheritance of pseudocholinesterase: one normal, two atypical, and one silent allelic gene. They form pseudocholinesterase with varying activity. Quantitative reduction of pseudocholinesterase occurs in patients who receive organophosphorus compounds like the antineoplastics agents cyclophosphamide and thiotepa (N, N', N''-triethylenethiosphosphoramide). Ecothiopate iodide, an anticholinesterase miotic used in the treatment of glaucoma, also reduces pseudocholinesterase levels and Trasylol, a polypeptide inactivator of kallikrein, obtained from animal sources, is also reported to cause prolonged apnoea.

Porphyria

The porphyrias are a group of diseases, mainly hereditary in origin, which have many different symptoms. In some the only problem is an undue sensitivity to sunlight; in others the normal life of the patient may be shattered by devastating attacks of abdominal pain, paralysis of limbs and profound mental upset. These diseases have one thing in common—a marked disturbance of porphyrin metabolism, of which the striking clinical impression is often the passage of dark-red urine. The porphyrins are highly-coloured red-pigment substances which are related to the synthetic pathways used in the production of haem, and hence of haemoglobin, myoglobin and the cytochromes. The steps in this pathway are shown in Table 5.1. This synthesis can probably occur in cells generally, but the most important organs in which it takes place are the bone marrow and liver. The porphyrias involve as a primary or secondary biochemical abnormality, a fault in the synthesis of porphyrins with increased urinary and faecal excretion of metabolic intermediates.

The two commonest porphyrias in Britain are acute intermittent porphyria (A.I.P.) and porphyria cutanea tarda (P.C.T.), one of the cutaneous hepatic porphyrias.

In A.I.P. the urine contains very high levels of delta-aminolaevulinic acid (ALA) and porphobilinogen, both porphyrin precursors; while in P.C.T. the urine and plasma contain uroporphyrin with no increase in the precursors. (The excess uroporphyrin found in the urine in A.I.P. is formed non-enzymatically in the acid milieu of the urine.)

It is thought that the high circulating plasma levels of uroporphyrin in P.C.T. contribute to the photosensitivity. Patients with A.I.P. have in their livers a striking increase in ALA synthetase, the rate-limiting enzyme in porphyrin synthesis. There has been controversy as to the level of this enzyme in the livers of patients with P.C.T., but the overall evidence suggests that it is raised in this condition also. It seemed likely that some additional biochemical abnormality is present in A.I.P., preventing the excessive formation of porphyrins in the liver, which always occurs in P.C.T. Work has lately been published which gives a rationale for this chemical finding in A.I.P. Uroporphyrinogen I synthetase (urosynthetase) and uroporphyrinogen III co-synthetase are two enzymes which catalyse the conversion of 4 molecules of porphobilinogen to uroporphyrinogen III. In A.I.P. there is decreased activity of urosynthetase in the liver, and it has been suggested that the partial block at this level not only would interfere with the negative-feedback regulation of hepatic ALA synthetase, leading to overproduction of ALA and porphobilinogen, but also would account for the normal or only slightly increased urinary and faecal porphyrins.

The cutaneous porphyrias. The first description of cutaneous porphyria (porphyria cutanea tarda, P.C.T.) was made in Glasgow at the end of the nineteenth century in two brothers who were fishermen from Stornoway. These brothers were extremely sensitive to sunlight, and each summer they suffered itching and burning sensations of the hands and face, followed by blistering. Their urine was burgundy red in colour and contained a great excess of porphyrins.

Skin photosensitivity and a marked increase in porphyrin formation by the liver can occur as a result of certain drugs (see Table 5.2) and other toxic substances. In South Africa many thousands of Bantus have this type of porphyria, with evidence of hepatic dysfunction. The main precipitatory factor seems to be the drinking of adulterated alcohol brewed and sold in urban areas. In 1956 many thousands of cases of cutaneous porphyria occurred in south-east Turkey, especially among children below the age of 15 years. This was caused by the distribution of seed wheat (used for bread making) which was dressed with the fungicide hexachlorobenzene. Many children died and had evidence of severe mutilation of exposed skin on the face, legs and hands. They all had very large quantities of porphyrins in their urine and blood.

The purely cutaneous form of the disease (P.C.T.) has a hereditary component, but also occurs secondary to hepatocellular disease.

Acute porphyrias. The symptoms of acute porphyria are severe attacks of colicky abdominal pain, vomiting and constipation. There may be weakness or even paralysis of the limbs, and more rarely of the respiratory muscles. The psychiatric manifestations are most important, and the patient may be wrongly diagnosed as having hysteria, psychoneurosis,

DRUG TOXICITY

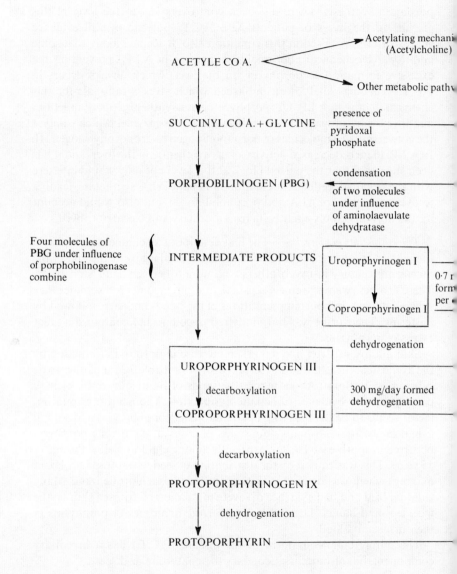

Table 5.1 Overall scheme of porphyrin synthesis

DRUG TOXICITY

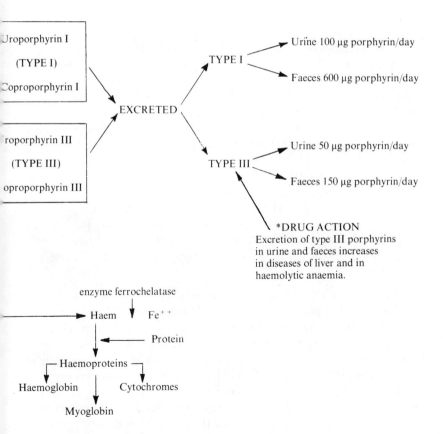

Table 5.2 Drugs that induce acute porphyria (A) or cutaneous porphyria (C) in man
(D'Arcy and Griffin 1972)

sulphonal	A and C
trional	A
tetronal	A
sulphonamides	A
amidopyrine	A
barbiturates	A and C
chlordiazepoxide	A
dichloralphenazone	A
meprobamate	A
phenytoin	A
tolbutamide	A and C
chlorpropamide	A and C
griseofulvin	A
oestrogens, progestogens and androgens including oral contraceptives	A and C
chloroquine	C

paranoia or schizophrenia.

The most important cause of the disease is the hereditary factor, since it is transmitted as a Mendelian dominant but with a varying degree of penetrance. The disease may be entirely symptomless or latent, and provoked only by the administration of drugs such as barbiturates, oral contraceptive agents or alcohol (see Table 5.2).

The causal relationship between some drugs and this disease has been known for many years. In 1889 Stockvis described the case of an elderly women who, after taking the hypnotic sulphonal, passed dark-red urine containing a pigment similar to haematoporphyrin, and who subsequently died. One year later, Harley reported the case of a patient who, after ingestion of sulphonal, exhibited many of the neurological features of what is now known as A.I.P. Several other cases of acute porphyria were described in the following years, and in some of these sulphonal, or the related drugs tetronal and trional, had been taken before the onset of symptoms.

Mixed porphyrias. These are a group of porphyrias in which there are symptoms of skin photosensitivity, and also features of acute porphyria with pain, vomiting, constipation, paralysis and psychiatric complications. This type of porphyria, also known as "variegate porphyria", is most common in South Africa, where there are about 10 000 cases, practically all of whom trace their descent to two Dutch settlers who married in the Cape in 1688. In the intervening period these porphyric families were just as large and prolific as other families, but it was the introduction of barbiturates at the beginning of the twentieth century that caused the disease to become much more lethal. In terms of practical therapy in patients with any form of porphyria, aspirin, methadone, pethidine and morphine are acceptable analgesics, but pentazocine and phenylbutazone should be avoided. The hypnotics of choice are chloral or triclofos. The following should not be given: barbiturates, dichloralphenazone glutethimide, meprobamate and benzodiazepines.

Drug-induced haemolytic anaemia in patients with inherited erythrocyte abnormalities
The life span of the erythrocyte in a normal subject is 110–125 days; in haemolytic anaemias the red cells are destroyed more rapidly, and the average life span of the erythrocyte is correspondingly shorter. It has been known for many years that when certain ordinarily harmless drugs are administered to some individuals in ordinary therapeutic doses, this is accompanied by rapid red-cell destruction and the development of acute haemolytic anaemia.

The most typical clinical manifestation of glucose-6 phosphate dehydrogenase deficiency is a haemolytic crisis precipitated by drugs (or by food or infections). Mild episodes may be recognized only by the incidental discovery of haemolytic anaemia in a patient receiving treatment for an unrelated condition. Neonatal hyperbilirubinaemia, leading to kernicterus, occurs in G6PD-deficient infants in Africa, China and the Mediterranean region after exposure of the child or mother to certain drugs or to naphthalene, and at times may occur even in the absence of such exposure.

The haemolytic drugs do not appear to have common chemical structural characteristics. The exact mechanisms of drug haemolysis are not fully understood, but they are related to defective metabolism of glutathione. Kidney and liver damage may increase blood levels of potentially toxic drugs, and thereby cause a relatively safe drug to be a highly haemolytic substance. Similarly, diabetic acidosis and other electrolyte disturbances may alter the haemolytic sensitivity of the red cell. The extent of haemolysis usually appears to be dose-related.

Haemolytic episodes may be precipitated by cutaneous contact with an offending agent, or by inhalation of its vapour (e.g. naphthalene) without actual ingestion.

The initial work identifying the mechanism of certain drug-induced haemolytic anaemias with deficiency of the enzyme G6PD was conducted in the US Army Malaria Research Unit at Stateville Penitentiary, Illinois, where the mechanism of the haemolytic anaemia induced by the antimalarial drug primaquine was investigated.

When a daily dose of primaquine (30 mg) is given to a sensitive individual, there is no untoward reaction for about 3 days. Then the urine begins to darken, the haemoglobin level, red-cell count and haematocrit fall. In the more-severe type of reaction, immediate transfusion may become necessary. If, in spite of these reactions primaquine is continued, the haemolytic symptoms subside after about a week. After about a month, red-cell production has speeded up and the disease is "self-limiting".

If primaquine is stopped for a month or two and then resumed, a new haemolytic episode similar to the initial one occurs. Once a self-limiting stage has been reached, an increase in dose may induce further haemolysis for a period until a new self-limiting phase is achieved.

The explanation for the haemolysis followed by recovery was demonstrated using red cells tagged with radioactive Fe^{59}. By administering radioactive iron Fe^{59} for a short period, newly-produced red cells marked with this tracer were produced in susceptible individuals. A two-week course of primaquine given to these subjects produced the expected haemolytic episode, but the young labelled red cells (8–21 days of age) were not destroyed. However, if primaquine was given later, when the labelled red cells were 55 days old, all the labelled red cells were destroyed.

It was demonstrated that the haemolytic episodes occurred only in subjects whose red cells were deficient in the enzyme G6PD, and that this deficiency in G6PD was more marked as the red blood cells aged.

In those countries where malaria is endemic, prophylaxis of malaria and treatment of the condition are an everyday problem. Knowledge of the incidence of G6PD deficiency in these populations was determined in epidemiological studies and is shown in Table 5.3.

Table 5.3 Incidence of glucose–6–phosphate dehydrogenase (G6PD) deficiency in Arab and African subjects

Country	Incidence per cent
Congo	3–30
East Africa and Tanganyika	2–28
Gambia	15
Ghana	24
Nigeria	6–17
Northern Nigeria	21
Palestine (Arab)	3.4
S. Africa (Bantu)	3.0
Sudan (Arab) Adults	8.1
New born (umbilical blood)	7.3

It was also found that G6PD deficiency existed in various forms, that a wide range of drugs could initiate haemolytic episodes in G6PD-deficient subjects (see Table 5.4), and that the severity of the haemolytic response was affected by the severity of the enzyme deficiency.

African type (A^-). The A^- type of G6PD deficiency is characterized by mild enzyme deficiency (mean enzyme activity 8–20 percent of normal) and high electrophoretic mobility. The youngest red cells have normal or almost normal enzyme levels, and red cells younger than 50 days have sufficient enzyme activity to be protected against damage by haemolytic drugs. Thus, only the old cells are susceptible to destruction and, even if the offending drug continues to be administered, haemolysis will be self-limited. The risk of potentially fatal haemolysis is therefore less than with the Mediterranean type of enzyme deficiency, and fever drugs are potentially toxic (Table 5.4).

Table 5.4 Drugs reported to induce haemolysis in subjects with G6PD deficiency

Drug	Haemolysis	
	Negro subjects	Caucasian subjects
acetanilide	+ + +	
dapsone	+ +	+ + +
furazolidone	+ +	
furaltadone	+ +	
nitrofural	+ + + +	
nitrofurantoin	+ +	+ +
sulfanilamide	+ + +	
sulfapyridine	+ + +	+ + +
sulfacetamide	+ +	
salazosulfapyridine	+ + +	
sulfamethoxypyridazine	+ +	
thiazosulfone	+ +	
quinidine		+ +
primaquine	+ + +	+ + +
pamaquine	+ + + +	
pentaquine	+ + +	
quinocide	+ + +	+ +
naphthalene	+ + +	+ + +
neoarsphenamine	+ +	
phenylhydrazine	+ + +	
toluidine blue	+ + + +	
trinitrotoluene		+ + +

Mediterranean type. The Mediterranean type of G6PD deficiency is characterized by severe enzyme deficiency (0–4 percent enzyme activity). Identification requires further enzyme characterization. Enzyme deficiency affects even the younger red cells, and haemolytic episodes are therefore not self-limited. Haemolysis is thus more severe and more often life-threatening, and more drugs are potentially harmful (Table 5.4).

Other common types of severe G6PD deficiency. The variants of G6PD deficiency that are common in East and South-East Asia (e.g. the variants

Canton and Union) differ from the Mediterranean and A⁻ types. The changes in enzyme activity and the clinical implications with regard to haemolysis have not yet been determined with these variants. It is likely that some at least may be as severe as the Mediterranean types of deficiency.

Much more work is required to characterize these G6PD types and to assess the susceptibility to haemolysis of red cells containing them before their public-health importance can be fully evaluated.

Until more information becomes available, variants that have not yet been defined should be considered a potential risk if they are associated with an enzyme activity of less than 10 percent. It should be stressed, however, that there is no complete correlation between enzyme activity, as estimated by standardized *in vitro* techniques, and clinical severity.

Glucose-6-phosphate dehydrogenase deficiency is known to be a sex-linked trait. In the male, the gene is associated with the X-chromosome derived from the mother. Males with G6PD deficiency have a single enzyme-deficient red-cell population, and homozygous females also have a single red-cell population. Heterozygous females are more frequently encountered in a population than deficient males, and they have two red-cell populations, one normal and one deficient. The ratio of the two populations varies widely, with a mode of 50:50; in rare cases the proportion of abnormal cells may be as low as 1 percent or as high as 99 percent. Only abnormal cells are drug-sensitive, and in most heterozygous females drug-induced haemolysis is mild, since in the typical heterozygous female only half of the cells are enzyme-deficient. Only about one-third of all heterozygous females have a high enough proportion of abnormal cells to predispose them to clinically significant haemolysis.

Instances of haemolytic anaemia following administration of customary doses of these drugs are not limited to patients with inherited G6PD deficiency, but also apply to other inherited red-cell abnormalities. Haemolytic anaemia may develop on exposure to oxidant drugs in persons who, for example, inherit erythrocyte glutathione (GSH) reductase deficiency, erythrocyte GSH deficiency, or the haemoglobinopathy associated with haemoglobin Zurich; haemoglobin-H disease and possibly other unstable haemoglobins also may be implicated.

Participation of genetic factors in thromboembolism in women on oral contraceptive therapy
There is strong evidence that the risk of venous thromboembolism in association with the use of oral contraceptives is only about one-third as great among women who belong to blood group O as among those who belong to the other three groups.

There are as yet no available comparable data for cerebral thrombosis in women of blood group O on oral contraceptive medication and, at this stage, there is no reason to believe that the same genetic factors affect the risk of this condition which involves arterial rather than venous thrombosis. It is, however, obviously a study worthy of investigation.

It is interesting to speculate whether this reduced risk of venous thromboembolism in blood group O women on "the pill" can be explained by the relatively lower level of antihaemophilic globulin (factor VIII) in these women. Factor VIII is increased during contraceptive medication, and it may well be that due to its initially lower level in these women, the influence of oral contraceptives does not raise the level to that necessary for participation in, or contribution towards, thromboembolic episodes. Subjects of blood group O have a lower level of antihaemophilic globulin than people of blood group A, and it has also been shown that people of blood group O are more likely to bleed as a result of peptic ulceration than those of A, B or AB.

A prospective drug-surveillance programme, conducted in three Boston Hospitals, revealed a deficit of patients of blood group O among those who received anticoagulants for venous thrombosis. This observation led to a

Table 5.5 Relative risk of thromboembolism A/O and (A+B+AB)/O in each series of patients

Type of patient with thromboembolism	Comparison	Pool Estimated Risk from USA, Sweden, UK
Non-pregnant women not using oral contraceptives	A/O (A+B+AB)/O	1·8 1·8
Pregnant or puerperal women	A/O (A+B+AB)/O	2·1 2·1
Women using oral contraceptives	A/O (A+B+AB)/O	3·2 3·3
Control series	A/O (A+B+AB)/O	0·9 1·2

There was a conspicuous deficit of patients of blood group O in all the groups studied. This deficit was particularly prominent when the thromboembolism was associated with oral contraceptives or pregnancy.

co-operative study in the United States, Sweden and the United Kingdom to obtain the blood groupings of young women who developed venous thrombosis while taking oral contraceptives, or during pregnancy, or the puerperium, or at other times. These findings have been published by Jick and colleagues (1969) and are summarized in Table 5.5.

Regional factors in adverse reactions to drugs
Colitis is a known complication of lincomycin and clindamycin antibiotic

therapy, but the incidence of the condition varies enormously in different populations, from less than one in a thousand patients to as high as one in twenty.

The psychotropic agent clozapine was associated with a high incidence of agranulocytosis in Finland.

The most dramatic of these regional outbreaks of iatrogenic disease is the condition now widely referred to as S.M.O.N. (subacute myelo-optic neuropathy) which is generally believed to be due to isochlorhydroxyquinoline. The high incidence of subacute myelo-optic neuritis in Japan led to the establishment of a S.M.O.N. Research Commission in 1967. A nation-wide survey published by Kono (1971) revealed 7856 cases, of which 5048 were confirmed.

In most cases there is a history of chronic abdominal symptoms, usually of pain and diarrhoea preceding the neurological illness. There was usually a prodomal exacerbation of the abdominal symptoms before the acute or subacute onset of sensory neuropathy in the lower limbs. In about two-thirds of the patients there were painful dysaesthesia, and they developed atoxic gait; half the patients developed muscular weakness, particularly of the lower limbs. A quarter of the patients had visual impairment; in some patients there was complete optic atrophy.

In a survey of 1092 patients with S.M.O.N. a clear association with clioquinol ingestion was found in 944 (86·4 percent). The average total dose of clioquinol in the six months before the onset of S.M.O.N. was 40·1 g. There was a relation between the total dose, the severity of visual impairment, and the greenish discoloration of the tongue. In the green pigmentation of the tongue, clioquinol derivatives had been identified.

In the fatal cases of S.M.O.N. an average clioquinol intake of 141 g was estimated. The peak incidence of S.M.O.N. occurred in 1969 and fell very rapidly after September 1969 when a total ban on sale of clioquinol was imposed in Japan, where there were 184 clioquinol-containing products on the market.

Cases of S.M.O.N. have been reported from Australia, UK, Germany, Norway and the USA, but only in very small numbers, and they have also shown on association with clioquinol-containing preparations, usually Entero Vioform.

The reason for the enormous incidence of S.M.O.N. in Japan is unknown, but associated genetic, environmental or infective, possibly viral, factors have all been postulated.

Age factors
There are well-defined examples of age being a factor in drug metabolism which has a direct bearing on incidence of iatrogenic disease.

The newborn have a relatively lower glomerular filtration rate and renal plasma flow than adults, and are also seriously deficient in drug-metabolizing enzymes for at least the first month after birth. The latter is particularly marked in their failure of glucuronation. All these deficiencies are enhanced in premature babies; thus neonates may fail to metabolize effectively vitamin K analogues, sulphonamides, chloramphenicol, barbiturates, morphine and curare. Penicillin excretion is also delayed, although this may be a useful rather than harmful defect of renal function, since effective blood levels are maintained for longer periods.

In the geriatric patient, drug overdosage is particularly likely to occur if the drug remains active in the body until it is excreted by the kidneys. This is because renal function (glomerular filtration and tubular function) diminishes with increasing age, even in the absence of clinically detectable disease. A reduction of about 30 percent in the glomerular filtration rate and tubular function has been demonstrated in otherwise normal patients over 65 years of age, when compared with normal young adults. Indeed at 90 years of age the functional capacity of the "normal" kidney may be only half what it was at 30 years. Diminished renal function may be made worse by dehydration, congestive heart failure, urinary retention or diabetic nephropathy, all of which are more frequent in the elderly patient. Under these circumstances drugs such as streptomycin, digitalis and oral hypoglycaemic agents may prove a real hazard.

Chlorpropamide has a long action (half-life 40 hours) even in the presence of normal kidney function, but elderly women are particularly prone to risk of hypoglycaemic episodes with this long-acting agent, because of generally poor renal function and their greater liability to urinary infection. Under such circumstances, therefore, tolbutamide is safer, since it is inactivated by the liver and therefore has a much shorter half-life (about 5 hours). Even using tolbutamide in these elderly patients, it should be borne in mind that use of other drugs, particularly phenylbutazone and sulphonamides, can prolong the half-life of tolbutamide to as much as 17 hours.

In the elderly, if a night sedative proves necessary, barbiturates are best avoided altogether, since they increase nocturnal restlessness and often produce sufficient hangover to make the old person inactive, inattentive, and mentally confused and unsteady on his feet next day. Safer alternatives are chloral hydrate, nitrazepam or glutethimide.

The phenothiazine group of drugs has proved invaluable in geriatric practice by decreasing anxiety, hallucinations and delusions, but with these drugs, too, there are special hazards. Compared with younger subjects on phenothiazines, elderly patients are more liable to suffer from phenothiazine-induced parkinsonism. Because the phenothiazines lower body temperature, they contribute to the problem of accidental hypothermia in the elderly, more especially if there is any reduction in thyroid function. Hypotension is also an important effect of phenothiazines in the elderly, causing increased risk of dizziness and falls. In those geriatric patients who develop cholestatic jaundice due to phenothiazines, their rate of recovery of liver function is greatly impaired.

The toxic effects of digitalis on the heart are essentially an extension of its therapeutic effects. The age of the patient is a predisposing factor in the development of toxic manifestations of digitalis. The elderly patient, whose heart is usually more severely damaged, is more likely to develop serious toxic manifestations. Indeed the incidence of digitalis intoxication increases with age out of proportion to the increasing incidence of heart disease.

Disease states

Renal failure. The action of a drug can be prolonged or increased in the presence of renal failure; this can be illustrated by the cumulative action of phenformin and the sulphonylureas, which have been described as causing profound hypoglycaemic episodes in the presence of renal failure. Toxic manifestations of a drug are more likely to occur in the presence of diminished excretion of the drug; indeed, in all the quoted cases of deafness attributed to ethacrynic acid and frusemide, this toxicity has occurred in patients with reduced renal function.

Liver disease. It is common knowledge that morphine may precipitate coma in patients with cirrhosis, and that paraldehyde causes profound sleep in some patients with liver disease. It is difficult to show that the half-life of these drugs is increased in patients with cirrhosis, though recent investigations indicate that this is probably so. In other words, failure to metabolize a drug due to liver failure can enhance and prolong the action of the drug.

Myasthenia gravis. A dual block of neuromuscular transmission following succinylcholine is seen in certain normal patients for no apparent reason, but this type of block is particularly prevalent in patients suffering from myasthenia gravis. This dual neuromuscular block consists of a depolarizing block followed by a non-depolarizing or curare-like block.

Gout and diabetes. Thiazide diuretics may cause diabetogenic or uricogenic effects even at normal therapeutic dosage. These unwanted effects occur in patients with a predisposition to these conditions. In addition, the newer diuretics chlorthalidone and frusemide may also precipitate gout in susceptible patients, and both these and the thiazide diuretics are contraindicated in severe renal and hepatic failure.

Abnormal eyes (shallow anterior chamber). In patients with a shallow anterior chamber of the eye, dilatation of the pupil with mydriatic drugs causes the iris to block the outflow of the aqueous humour, and angle closure glaucoma results. This occurrence is more likely in the elderly than in the younger patient. In addition the eyes of mongols are more sensitive to mydriatic drugs than those of normal patients.

Hypersensitive and allergic reactions to drugs
Drug hypersensitivity or allergy is a predictable phenomenon in terms of assessing the hazard for any particular or individual patient. Although it is known that some drugs are more allergenic than others, this information is based on clinical evidence of the incidence of allergic reaction, and animal toxicity testing is virtually useless in assessing risks of this type.

In hypersensitivity reactions there is usually no reaction on first exposure (i.e. during the first few days of treatment) unless there has been previous exposure (and sensitization) to a similar chemical agent. Thus neomycin may cause topical skin hypersensitivity, but because of certain common chemical groupings, subsequent treatment of the patient with streptomycin or kanamycin may lead to a generalized reaction. Strangely this cross-sensitivity principle does not necessarily extend to stereoisomers, e.g. quinidine and quinine may not cross-react.

Drug-induced allergic skin reactions
Skin reactions to drugs probably present the commonest and best-known examples of allergic reactions to drugs, and may occur following topical application to the skin or following systemic administration.

Allergic contact dermatitis. Allergic contact dermatitis occurs when a patient becomes sensitized to a substance in a topically applied medicament. The sensitizer may be an ointment base, such as lanolin; a preservative, such as parabens or ethylenediamine; or an active ingredient, such as neomycin. The sensitivity may be masked if the preparation also contains a corticosteroid, so that the condition under treatment worsens insidiously. Fourteen percent of 4000 consecutive patients with eczema from 5 European Skin Clinics tested recently by the International Contact Dermatitis Research Group were considered to have allergic contact dermatitis due to medicaments, neomycin and benzocaine being the most frequent sensitizers (4 percent each). They also found that 40 percent of women with dermatitis of the lower leg had developed contact dermatitis due to medicaments.

It seems that patients with atopic eczema are no more likely to develop contact dermatitis than those patients with other forms of endogenous eczema. But any patient with long-standing eczema of any sort is likely to have had prolonged treatment with medicaments, and therefore the opportunity of becoming sensitized. Therefore, when treating patients with topical remedies, it is important to discover whether or not there is a past history of allergic reactions, especially to chloramphenicol and other antibiotics, because latent hypersensitivity from past episodes of contact dermatitis due to these substances can persist from many years. Cross-sensitivity can also occur, e.g. patients sensitive to neomycin also cross-react to framycetin and gentamicin.

The chances of developing allergic contact dermatitis vary with the site of application. Thus, the application of drugs to the lower limbs when stasis and ulceration are present carries special risk; systemic absorption can occur and may result in immediate-type hypersensitivity with anaphylaxis.

Contact eczema of the eyelids can result both from ointments and from eyedrops. Atropine is still commonly used; sensitivity is not rare and can be proved by patch testing. Amethocaine eyedrops used for surgical procedures can cause severe sensitivity. Sulphacetamide occasionally causes contact dermatitis, but it is not severe in contrast to the reactions with chloramphenicol, which are both severe and by no means infrequent). Chloramphenicol should only be used where there are strict bacteriological indications. Neomycin sensitivity is not uncommon, and sensitivity to benzalkonium is known, although probably rare.

It is important to appreciate that when a patient's skin has been sensitized to a drug and that drug, or one closely related to it chemically, is given systemically, a skin eruption may occur. This may resemble the original contact dermatitis, but is usually more widespread.

Contact sensitivity dermatitis can be serious for medical and nursing personnel, because desensitization is not possible. Nurses are at special risk when solutions, rather than tablets, of chlorpromazine are used, and when injections are often employed; and a similar problem exists with streptomycin, although these dangers can be minimized by careful injection technique. Workers in renal dialysis units are at risk from formalin sensitivity. Doctors and nurses may become sensitive to cleansing agents such as cetrimide. Rubber-glove sensitivity is a severe handicap to those who need them, especially in operating theatres; but fortunately alternative gloves containing different rubber chemicals are now available. Sensitivity to orthopaedic acrylic cement, which can penetrate rubber gloves, is a hazard, especially to theatre sisters and surgeons. Dentists can become sensitized to amethocaine, and less commonly to other materials, such as mercury and chromic acid, which may also affect dental technicians.

Skin rashes from systemically administered drugs. The penicillins are the commonest drugs to cause rashes, and indeed may be responsible for a very high proportion of all adverse reactions. Evidence imputes immune mechanisms in their pathogenesis, whether from the "naturally" produced penicillins (e.g. benzylpenicillin and procaine penicillin) or the semi-synthetic group (e.g. ampicillin and cloxacillin). Mostly the rashes must be considered as cutaneous markers of a generalized hypersensitivity state.

The reported incidence of rashes varies enormously. It depends upon many factors: whether the firgures were sought prospectively or retrospectively; whether they were collected personally or reported "officially", which often means haphazardly; for what disease or microorganism the drug was given; and the prescribing habits of the doctors as regards dosage, length of treatment, route of administration and so on.

In a small prospective study, toxic erythema or urticaria was seen in 15 patients of 32 receiving ampicillin for definite infections in the lungs (often *Haemophilus*) or following gut or urinary tract operations (usually Gram-negative organisms). Most of the patients were symptomless and, since the rash was transient and often very sparse, it was detected only by examining the patients regularly on alternate days. The main signs are usually seen 8–12 days after starting the drug, but a fleeting rash may be seen on the third to fifth day, and is usually ignored by both patient and doctor.

Anaphylactic reactions to the penicillins kill an appreciable number of patients every year. Although the incidence of such a disaster may be perhaps between one in 10 000 and one in 100 000 patients in a population not suspected of penicillin sensitivity, the drugs are so widely prescribed that this is an event that can never be ignored. Immediate urticarial reactions severe enough to be life-threatening are probably four times as common. Nevertheless these are relatively rare events in comparison with the exceedingly common "ordinary" urticarias and the exanthema that is morbilliform-toxic erythema and erythema multiforme.

The antigen is seldom the drug in its chemically pure form, but a degradation product either on its own or as a protein complex. Metabolic degradation or biotransformation in the patient or spontaneously occurring during the manufacture and storage of the drug is responsible for the formation of the chemically reactive haptenic compounds. Also, in the case of antibiotics particularly, traces of contaminants carried over from the complex fermentation processes and involving microorganisms cultured on special media have to be considered. Such matters must be partially responsible for variable figures quoted for the incidence of eruptions. Ampicillin, for example, contains impurities which may polymerize while in solution, depending upon its concentration, and may become degraded during its metabolism. It is disputed by others that ampicillin always contains protein impurities and that the non-protein components are antigenic. Certainly "purification" of ampicillin reduces the incidence of eruptions.

Penicillins contain a number of antigenic determinants with high sensitizing potential. Allergic penicillin rashes can be discussed in relation to the diagnostic skin tests that detect IgE antibodies using two groups of antigen:

1. The major haptenic grouping—the penicilloyl derivative of penicillin acid, probably responsible for the serum sickness type of reaction.
2. The quantitatively minor (5 percent) determinants that include penicillin breakdown products, ill-fated and variable in nature, probably responsible for the immediate-type anaphylactic reactions.

In the case of ampicillin, the immediate type of urticarial reaction, as seen with the natural penicillins, is less frequent than the toxic erythema. The latter indeed may represent an "ampicillin specific" reaction not involving the penicilloyl group derived from the 6-aminopenicillanic acid nucleus of the natural penicillins. Cross-hypersensitivity, therefore, between the natural and the semisynthetic penicillins is not invariable. Indeed, some investigators tend to consider on the basis of skin-testing results that the exanthem associated with benzylpenicillin therapy may not have a true immunological basis, whereas others consider that it is a manifestation of a delayed-type hypersensitivity reaction.

Nearly every patient with infectious mononucleosis treated with ampicillin develops a toxic erythema, in contradistinction to those treated with a natural penicillin. It has been stated that "patient with sore throat + ampicillin rash = infectious mononucleosis until proved otherwise."

Patients with gout given allopurinol (Zyloric) for the treatment of their raised blood uric acid have a higher-than-normal incidence of allergic reactions to ampicillin, but it is not clear why this is so.

Allergic contact dermatitis in nurses handling ampicillin is a real danger and suggests that the practice of applying ampicillin to surgical wounds is undesirable. The signs of allergic dermatitis from penicillin may recur following subsequent systemic administration.

Drug-induced allergic asthma

Asthma with or without generalized anaphylactic shock or urticaria is probably mediated by Type-I (reaginic) allergic hypersensitivity. Type-I hypersensitivity has been reported following the administration of the penicillins and tetracylines, erythromycin, neomycin, streptomycin, griseofulvin, cephaloridine, ethionamide, monoamine oxidase inhibitors, radio-opaque organic iodides, local anaesthetics, mercurials, vitamin K, bromsulphthalein, suxamethonium, antisera, vaccines and allergenic extracts.

Asthma may also occur as part of a drug-induced serum sickness syndrome, the symptoms of which include fever, arthralgia, urticaria, maculopapular eruptions and lymphadenopathy; Type-III hypersensitivity (Arthus type) may be involved in the mechanism of this syndrome.

Though the mechanisms of most drug-induced asthmas are poorly understood, the comparatively rare pituitary-snuff-taker's lung is one in which the mechanism is known. This syndrome consists of an extrinsic allergic alveolitic in which patients have dyspnoea with cough and widespread crepitations. Chest radiographs show extensive miliary mottling. Microscopically, the lungs show fibrosis in alveolar walls with collections of lymphocytes, plasma cells and scanty eosinophils. There is also an intr-alveolar exudate and formation of a hyaline membrane. The presence in these patients' sera of precipitins against bovine and porcine serum proteins and pituitary antigens suggests that this lung reaction is mediated by a Type-III hypersensitivity. In addition, the inhalation of pituitary snuff used in the treatment of diabetes insipidis may provoke rhinitis (Type-I hypersensivity) and allergic asthma (Type-I and Type-III hypersensitivity).

The commonest cause of drug-induced asthma is aspirin. Asthmatics who are sensitive to aspirin constitute a characteristic clinical group. They have generally suffered from rhinitis and nasal polyposis for several years before the onset of their asthma. The asthma usually starts in middle life and tends to become chronic shortly after its onset. Asthma symptoms precede the development of aspirin intolerance by months or years, though it is rare that aspirin itself causes the first asthmatic attack. Angioneurotic oedema and urticaria may also accompany these aspirin-induced asthma attacks. Aspirin-induced asthma is characterized by the onset of symptoms 20–30 minutes after the ingestion of the drug; it is usually severe and prolonged, and occasionally fatal. Some workers think that patients with aspirin-induced asthma have a poor prognosis; however, long-term therapy with corticosteroids may control their disease.

Allergic drug-induced haemolytic anaemia

A number of drugs have been incriminated in haemolytic processes mediated by antibodies. These haemolytic anaemias are associated with a positive direct Coombs test, but at least three different mechanisms are involved. These can be distinguished, in part, by means of the modification of the Coombs antiglobulin technique that recognizes two major types of red-cell coating with proteins. One is the antigamma reaction that recognizes the immunogloblin g (IgG). The second detects the presence of the complement (c) system on the surface of the red cell. This is the anti-non gamma or anti-C' reaction.

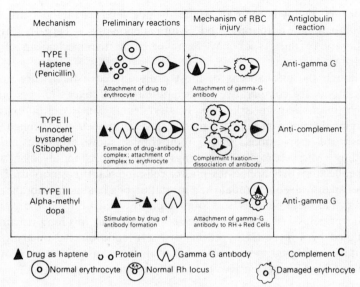

Figure 5.2 Diagrammatic representations of types of drug-induced Coombs-reactive haemolytic amaemias (redrawn from Wintrobe, 1969).

Table 5.6 Drugs having the property to induce antibody-mediated haemolytic anaemias in susceptible subjects.

Class of Drug	Member
hypotensive	methyl dopa
antibiotic	penicillin
	nalidixic acid
chemotherapeutic agents in schistosomiasis	stibophen (sodium antimonyl pyrocatechol disulphonate)
cinchona alkaloids	quinine, quinidine
analgesic/antipyretic	phenacetin
	dipyrone
	mefenamic acid
anti-leprosy agents	dapsone
anti-tubercular agents	para aminosalicylic acid (PAS)
sulphonamides	sulpha salazine

The three types of drug-induced Coombs positive haemolytic anaemias have been termed the alpha methyl dopa type, the innocent bystander type, e.g. stibophen type, and the haptene type, e.g. penicillin. In figure 5.2 are illustrated the various mechanisms of antibody mechanisms responsible for producing haemolytic anaemia, and in Table 5.6 is given a list of drugs known to cause haemolytic anaemia.

The oculomucocutaneous syndrome of Practolol

Practolol was first marketed in the United Kingdom in June 1970 as a selective beta adrenergic blocking agent and was widely used in the treatment of angina and hypertension. In the autumn of 1974 it was estimated that some 250 000 patients were under treatment with Practolol. In May 1974 Felix and Ive reported the association of a psoriaform eruption associated with Practolol, and in June 1974 Wright reported the occurrence of eye changes and psoriaform skin lesions with Practolol. In December 1974 Brown *et al.* published reports on three patients who developed sclerosing peritonitis while receiving long-term Practolol therapy.

The clinical syndrome has now been clearly defined and shown to be causally related to Practolol. The main features of the syndrome are now known to be a specific psoriaform rash, ocular lesions (xerosis/sclerosing conjunctivitis/keratitis), deafness caused by both secretory otitis media and cochlea damage, and sclerosing serositis, the main component of which is sclerosing peritonitis. The features of the syndrome may occur singly or in combination, and some 600 cases have been identified. The syndrome is thought to have an immunological basis. In the most severe cases of eye damage it is known that the patients have developed positive antinuclear factors, and in some an additional pemphigus-like intercellular antibody has been identified.

In 16 cases of severe Practolol-induced sclerosing peritonitis reported from one centre there were, however, no immunological changes.

DRUG REGULATORY AUTHORITIES

Attempts to minimize the hazards of adverse reactions to drugs have been tackled in most of the developed countries of the world by the establishment of some form or other of drug regulatory authority which assesses drugs with respect to their safety, quality and efficacy.

In the United Kingdom, drug safety was not the subject of any legislation until after the thalidomide disaster, following which a Joint Sub-Committee of the English and Scottish Standing Medical Advisory Committees recommended the establishment of an expert committee to review the evidence as to new drugs and to offer advice on their toxicity. The Committee on Safety of Drugs was set up in June 1963 by the Health Minister in consultation with the medical and pharmaceutical professions and the pharmaceutical industry. The Committee on Safety of Drugs had no legal powers, but worked within the voluntary agreement of the Association of the British Pharmaceutical Industry and the Proprietary Association of Great Britain. These two organizations promised that none of their members would put on clinical trial, or release for marketing, a new drug against the Committee's advice, which they undertook always to seek. The Medicines Act 1968 replaces most of the previous legislation on the control of medicines for both human and veterinary use. The Act is administered by the Health and Agriculture Ministers of the United Kingdom who constitute the Licensing Authority. In September 1971 the Committee on Safety of Drugs ceased to exist and was replaced by the Committee on Safety of Medicines, which acts as an advisory body to the Licensing Authority.

Application for clinical trial certificate

The clinical trial certificate application is usually made by the drug manufacturer. The application should state clearly the number of patients to be treated, the indications for which they are to be treated, the maximum daily dosage to be employed, and the duration for which dosage is to continue. In addition, details of trial design and safety monitoring in terms of haematological and serum biochemical studies, and urinalysis and frequency of clinical examination of the patients is expected. The application for the clinical trial certificate should be accompanied by experimental data. The pharmacological studies should demonstrate the full mode of action of the drug substance by the proposed clinical route of administration. These studies should give sufficient promise of therapeutic potential to justify the study.

Details of the metabolism of the drug in animals and man in terms of absorption, plasma half-life, rate of urinary and faecal elimination and, ideally, identification of urinary metabolites of the drug should be available. This is often the most difficult phase of the drug's evaluation in animals and man, and may be technically impossible at the time the application for a clinical trial certificate is made.

Data from repeat dose toxicity studies in at least two species is normally required; the dosing should have been conducted by the proposed clinical route of administration in man, and the duration of dosing should be appropriate to the proposed clinical use of the drug. These studies should be conducted with at least three dose levels; the lowest dose level should be in the therapeutic range, and the highest should have been selected after preliminary dose ranging studies to reveal the target organ. The purpose of toxicity tests is to demonstrate a pattern of toxicity to the clinical pharmacologist, not to provide a marketing organization with a testimonial establishing the safety of the product. Throughout the toxicity studies, repeated haematological and serum biochemical monitoring is essential, and at the completion of the study the animals should be autopsied, and histopathological examinations of the major tissues performed.

Reproduction studies should be conducted in such a manner as would reveal the effect of the drug on each of the following mechanisms of producing foetal abnormality, or foetal loss, or damage to the offspring in later life:

(i) The male and female gametes resulting in sterility or the formation of abnormal young.
(ii) Intrauterine homeostasis and nutrition of the foetus with particular respect to the processes of implantation and normal placental function.
(iii) Organogenesis.
(iv) Direct toxic effects on the foetus.
(v) Maternal metabolism producing secondary effects on the foetus.
(vi) Uterine growth or development.
(vii) The processes of parturition.
(viii) Post-natal development, suckling of the progeny and maternal lactation.
(ix) Late effects on the progeny.

These new standards for reproduction studies were introduced in the United Kingdom in January 1975.

Application for marketing or product licence

When clinical trials have been completed and a request is received for marketing, the assessment will be based largely on the clinical documentation in relation to the proposed clinical use. In the early clinical studies, evidence will have been obtained on the absorption, distribution, metabolism and excretion of the drug. In addition, the clinical pharmacology will have been studied, when appropriate.

The preparations of the drug proposed for marketing must show an adequate stability and uniformity of content. It must be clearly demonstrated that methods of formulation and preparation do not modify the drug action or interfere with its biological availability.

Evidence will be sought in the documentation for any sign of possible organ toxicity which might be revealed from the monitoring by haematological and clinical chemical methods. Problems of possible interactions with other drugs must also be scrutinized. In addition at this stage, if the drug is intended for prolonged use, further studies will be required of long-term toxicity in animals.

Finally, the proposed promotional literature is examined to ensure that in the view of the Committee no extravagant or misleading claims are made and that the necessary precautions and contra-indications are adequately expressed.

Assessment of a drug is made in relation to pharmaceutical quality of ingredients, and the efficacy of the drug in proposed clinical use. The likely toxicity due to pharmacological overdose can be determined to a large extent from animal studies, always provided that the metabolism of the drug in the species and strain of animals used resembles that of man. Subject to similar reservations, the inherent toxicity and teratogenic potential can be reasonably assessed in animal studies. Additional information on the potential hazards of a new drug can be obtained from the reports of adverse reactions experienced during clinical trials.

Where adverse reactions to drugs occur because of genetic patient abnormality, or allergic or idiosyncratic reaction, assessment of the potential hazards by extrapolating from animal studies is of little predictive value. In these instances, the hazard may become apparent only after marketing of the preparation and treatment of large numbers of patients, often for long periods, depending on the incidence of the particular abnormality in the population. It is for this reason that continual monitoring of adverse reactions on marketed drugs is necessary.

The surveillance of drugs in the United Kingdom after marketing is directed by the Adverse Reactions Sub-Committee of the Committee on Safety of Medicines. The primary mechanism for monitoring is based on a voluntary and spontaneous reporting system, using a simple reply-paid yellow card, whereby doctors and dentists are encouraged to report any suspected adverse reactions.

It is currently estimated that only 1–10% of all adverse reactions are reported by doctors. One of the major problems is that a doctor does not necessarily recognize set-backs or adversity in his patient's condition as being an adverse reaction to a drug, and so such a reaction is unreported. The most obvious example of this is the recent practolol-induced oculomucocutaneous syndrome. If the chances of detecting such dangers before they reach major proportions are to be improved, it is essential that a more sensitive system be introduced. A scheme that would achieve this end is described below; it has been considered by the Committee on Safety of Medicines, who have recommended that the scheme should go out to consultation. This scheme has been called "Recorded Release".

Patients treated with a drug on a recorded release should be identified and their clinical progress closely followed for any unexpected events, as well as clearly identified adverse reactions. It is envisaged that when these case histories are analysed, it will be possible to recognize many of these untoward events as drug-induced adverse reactions because of their association with exposure to the drug.

The system of recorded release hinges firstly on the identification of patients treated with a specified drug and secondly on the identification of the prescribing doctor. The only document which exists that contains all the relevant data is the prescription form FP10 (or EC10 in Scotland), i.e. (i) name of drug, (ii) name of patient, (iii) name of prescribing doctor. Therefore the scheme hinges on prescription sorting, using the Prescription Pricing Authority (PPA) to sort the prescriptions. The identified prescriptions would be photocopied and supplied to a monitoring centre. The monitoring centre would then follow up the progress of these patients through their family doctors.

The Committee on Review of Medicines has been established with the function of undertaking an assessment and review of the quality, safety and efficacy of all medicinal products. This will ensure that products that were on the market before the setting up of the Committee on Safety of Drugs or the Committee on Safety of Medicines will be subjected to scrutiny, that all licences granted for medicinal products will be subjected to scrutiny, and that all licences granted for medicinal products will also be subject to an ongoing review.

Conclusion

Absolute safety cannot exist; the more potent the remedy, the more may be the capacity for harm. It is a difficult road to travel, for the sick are not safeguarded if undue restriction and precaution impedes therapeutic advance. Constant care is required to ensure that no patient is deprived of any therapeutic advance, even at an early stage of development of a preparation if the need arises.

Awareness of the possible hazards of medication, and of possible interaction between drugs on the part of those who use them, and the identification of these sections of the population who may be at special risk can only result in better therapeutics, with benefit to the patient in terms of both safety and efficacy.

FURTHER READING

D'Arcy, P. F. and Griffin, J. P. (1972), *Iatrogenic Diseases*, Oxford Univ. Press.
Griffin, J. P. and D'Arcy, P. F. (1975), *A Manual of Adverse Drug Interactions*, John Wright & Sons Ltd., Bristol.
Herbst, A. C. and Ulfeloer, H., Poskanzer, D. C. (1971), "Adenocarcinoma of the vagina: Association of Maternal Stilboestrol therapy with tumour appearance in young women," *New Engl. J. Med.*, **284**, 878–881.
Hutchinson, J. (1887), "Arsenic cancer," *Brit. Med. J.*, **2**, 1280–1.
Jick, H., Slone, D., Westerholm, B., Inman, W. H. W., Vessey, M. P., Shapiro, S., Lewis, G. P.

and Worcester, J. (1969), "Venous thromboembolic disease and ABO blood type: A cooperative study," *Lancet,* **1,** 539–42.
Jick, H. (1974), "Drugs—remarkably non-toxic," *New Engl. J. Med.,* **291,** 824–828.
Lawson, D. H. (1974), "Adverse reactions to potassium chloride," *Quart. J.,* **171,** 433–440.
Melmon, M. K. (1971), "Preventable drug reactions, causes and cures," *N. Engl. J. Med.,* **284,** 1361–1367.
Ogilvie, R. I. and Ruedy, J. (1967), "Adverse drug reactions during hospitalisation," *Canad. Med. Ass. J.,* **97,** 1450–7.
Robson, A. O. and Jelliffe, A. M. (1963), "Medicinal arsenic poisoning and lung cancer," *Brit. Med. J.,* **2,** 207–209.
Smith, J. W., Seidle, L. G. and Cluff, L. E. (1966), "Studies in the epidemiology of adverse drug reactions," *Ann. Intem. Med.,* **65,** 629–40.
Sommers, S. C. and McManus, R. G. (1953), "Arsenical tumours of the skin and viscera," *Cancer,* **6,** 347–359.
Seidle, L. G., Thornton, G. F. and Smith, L. E. (1966), "Studies on the epidemiology of adverse drug reactions. II Reactions in patients on a General Medical Service," *Bull. Johns Hopk. Hosp.,* **119,** 299–315.
Williams, J. B. R., Griffin, J. P. and Parkins, A. (1975), "Effect of Concomitantly Administered Drugs on the Control of Long term Anticoagulant Therapy," *Quart. J. Med.,* **176,** 675–685.
World Health Organization (1973), Pharmaceogenetics Technical Report Series No. 524.

Index

activation analysis, 70
aflatoxins, 30
  in peanuts, 30
age factors in drug reactions, 114
allergic contact dermatitis, 117
aluminium smelters and atmospheric pollution, 6
amalgam, 78
analysis of food, 14–36
analysis of foodstuffs for compounds resulting from environmental pollution, 17
analysis of hair, 69
analysis of naturally formed compounds in food, 27
animals as biological indicators, 11
anticoagulants, adverse reactions from, 90
antidepressants, 100
antioxidants, 34
arsenic and cancer, 92
  and smoking, 75
  content of Napoleon's hair, 84
  in detergents, 77
arsenicals, toxicity, 91
Aspergillus, 32
aspirin and asthma, 121
asthma, drug-induced, 120
atmospheric pollution, 4
atomic absorption, 25, 73, 80

bacteria growing in agar, 51
beard hair, analysis of, 77
biological examination of water, 52
  factors in water analysis, 54
  indicators, 1
    advantages and limitations, 2
  methods of environmental monitoring, 1–13
biphenyl, 23
black spot disease, 7
Burns, Robert, 83

cadmium in food, 25
cancer, vaginal, and diethylstilboestrol, 94
carbon, activated, 65
  monoxide as a biological indicator, 2
cataract, 91
CEF (chick edema factor), 24

Charles II, 83
chemical analysis of water, 45
chick edema factor (CEF), 24
chloroform toxicity, 91
chromium in food, 26
'Clean Air Research Pack', 5
clioquinol, 114
Clyde River Purification Act 1972, 64
coagulation, 64
coliform organisms, 50
colorimetric analysis of water, 46
Committee on Safety of Drugs, 123
  Medicines, 124
contact dermatitis, 117
contamination of water, 38
control measurements for water quality, 41
  of water quality, 61
copper in food, 26
cottonseed products, 31
cystine, 68

DDT, 17, 23
dental practice, mercury hazards in, 78
detergents, arsenic in, 76
diabetes and drug reaction, 116
dichlorvos (Vapona), 20
dieldrin, 17
diethylstilboestrol and cancer, 94
digitalis, adverse reaction from, 88
digoxin, adverse reactions from, 90
dimethyl sulphoxide (DMSO) and cataract, 91
disease and environment, 82
  states, 116
diseases, water-borne, 56
disinfecting water, 65
dissolved solids and conductivity, 54
disulphide bridge, 68
diuresis in treatment of drug overdose, 99
diuretics, adverse reaction from, 88
drinking water standards, 57
drug absorption, 98
  allergy, 117
  interaction, toxicity due to, 96
  interactions, sometimes useful, 99
  metabolism, 98
  reactions in Australia, 89

reactions in USA, 87
safety, 123
toxicity, 87–127
  inherent, 91
  trials, 124
  usage in USA, 89
drugs, fast and slow acetylation, 101
drug-induced allergic skin reactions, 117
  asthma, 120
  haemolytic anaemia, 109, 121

ecological investigations of pollution, 11
electron-capture detector, 18
emission spectroscopy in water analysis, 48
Entero Vioform and subacute myelo-optic neuritis, 114
environment and disease, 82
  historical, 82
environmental lead, 80
Escherichia coli, 50

filtration, 64
fish as pollution indicators, 11
flameless atomic absorption, 75
fluorine as atmospheric pollutant, 5
  from aluminium smelters, 6
food additives, 34
  dyes, 34
fumigants in food, 20
fungi as biological indicators, 7
fungicides in food, 21
Fusarium, 33

gamma-ray spectroscopy, 72
gas chromatography, 18
genetic factors in drug toxicity, 101
gout and drug reaction, 116
gravimetric analysis of water, 49
Great Lakes, 62

haemolytic anaemia, drug-induced, 121
hair analysis, 69
  as a means of identification, 69
  as a mirror of the environment, 66–86
  follicles, 66
  growth, 66
    rate, 67
  Roman, 82
  structure, 66
hardness of water, 54
heavy metals, accumulation by mosses, 7
  in food, 24
  in water, 54
herbicides in food, 20
hollow-cathode lamp, 74

human water balance, 40
hypersensitive and allergic reactions in drugs, 117

iatrogenic disease, 89
indicator, biological, 1
  species, 1
industrial use of water, 59
insecticides, 17
  organochlorine, 3, 17
insulin, adverse reaction from, 88
intolerance of drugs, 100
invertebrate animals as pollution indicators, 11
ion-specific electrodes, 49
iron in food, 26
isoniazid, 101

keratin, 67

Lake Michigan, 62
Lake of Zurich, 63
lakes as water supply, 62
lead, environmental, 80
  in food, 25
  in hair, 80
    children's hair, 81
lichens as biological indicators, 4
  as indicators of radioactive fallout, 7
lindane, 17
liver disease and drug reaction, 116
London's water supply, 62

Medicines Act 1968, 124
membrane filtration, 51
mercury and arsenic in human head hair, 85
  hazards in dental practice, 78
  levels in dental workers, 79
  poisoning in Iraq, 81
metal binding in hair, 67
metals in hair, 67
methyl bromide, 20
microbiological analysis of water, 44
  examination of water, 50
  factors in water analysis, 55
monitoring of water quality, 59
mosses as biological indicators, 4, 7
Mowbray, Anne, 82
multiple ion monitoring, 33
multiple-tube test, 51
myasthenia gravis, 116
mycotoxins, 30, 32

Napoleon, 84
Nessleriser, 47

INDEX

neuromuscular blocking drugs, abnormal response to, 102
nitrosamines, 27

occupational poisoning by arsenic, 77
odour of water, 49
oral contraceptives, hazards, 112
organochlorine insecticide residues, 17
    pesticides, 3, 17
organoleptic factors in water analysis, 52
organophosphorus insecticides in food, 19
ozone, 8, 9

parathion, 19
PCBs (polychlorinated biphenyls), 23
Penicillium, 32
permanent waving, 68
pesticides, 17
    in food, 16
pH of water, 53
phenacetin and kidney damage, 96
phosphine, 21
photometer in water analysis, 47
physico-chemical factors in water analysis, 53
plasma protein binding, 98
pollution of water, 40
polychlorinated biphenyls (PCBs) in food, 23
polychlorodibenzo-p-dioxins, 24
polycyclic hydrocarbons in food, 33
porphyria, 104
    acute, 105, 108
    cutaneous, 105
    mixed, 108
porphyrin synthesis, 106–107
potassium choride, adverse reactions from, 90
Practolol, 123
public water supplies, 56

radioactivity in water, 49
rain water, 61
receptor sites, 99
recreational use of water, 57
renal excretion, 99
    failure and drug reaction, 116
reservoir, 64
river classification, 59
    pollution, 58
rivers as sewers, 41
    as water supply, 63
Roman hair, 82

signficance of water analysis, 52
silver in food, 25
skin rashes from drugs, 118
smoking and arsenic, 75
specific electrodes, 49
sulphur dioxide as atmospheric pollutant, 4
sulphydryl (SH) groups in hair, 68
Sweetheart Abbey, 82
synthesis of porphyrin, 106–107

tar spot disease, 8
taste of water, 49
thalidomide, 92–94
thromboembolism and oral conceptives, 112
tin in food, 27
tobacco plants as indicator of ozone, 10
toxic factors in water, 54
toxicity due to drug interaction, 96
trace element concentrations in whole body, blood and hair, 68
trace-metal analysis, 25
trace metals in hair, 68
    in water, 38
trees, effect of atmospheric pollution, 9
turbidity of water, 47
Turkey X disease, 30

underground water supplies, 63

vaginal cancer and diethylstilboestrol, 94
Vapona, 20
vascular plants as pollution indicators, 8
vinyl chloride in food, 27
viruses in water, 51
volumetric analysis of water, 45

water analyses at source, 45
water analysis, scope, 55
    significance, 52
water balance in man, 40
water cycle, 38
    control measurements, 41
    industrial use, 59
water quality control, 37–65
    monitoring, 59
    parameters influencing, 42
water, recreational use, 57
water supply, London, 62
    safe, provision of, 64
water-borne diseases, 37, 56
white hair, 67